AF559693

Thread

Thread

Naresh Grover

Thread

ISBN 978-93-5111-648-6

Published in 2015 in India by

RANDOM PUBLICATIONS

4376-A/4B, Gali Murari Lal, Ansari Road
New Delhi-110 002
Phone : +9111-43580356, 011-23289044, 011-43142548
e-mail: sales@randompublications.com,
info@randompublications.com, randomexports@gmail.com

Reprinted 2026

Type Setting by : Friends Media, Delhi-110089

Preface

Textiles have such an important bearing on our daily lives that everyone should know something about the basics of fibres and their properties. Textile fibres are used for a wide range of applications such as covering, warmth, personal adornment and even to display personal wealth. Textile technology has come a long way in meeting these requirements. A basic knowledge of textile fibres will facilitate an intelligent appraisal of fibre brands and types and help in identifying the right quality for the application. Thread, tightly twisted ply yarn having a circular cross section and used in commercial and home sewing machines and for hand sewing. Thread is usually wound on spools, with thread size, or degree of fineness, indicated on the spool end. Cotton thread is compatible with fabrics made from yarn of plant origin, such as cotton and linen, and for rayon, made from cellulose, a plant substance. Silk is suitable for silks and wools, both of animal origin; and nylon and polyester are appropriate for synthetics and for knits having a high degree of stretch.

I would like to thank my team for standing beside me throughout my career and writing this book. My special thanks go to "Random Publications" who have published the book.

– Naresh Grover

Contents

1

Introduction

THREAD ESSENTIALS

Aside from fabric, thread is perhaps the most essential component of sewing—yet we often sew without giving much consideration to our thread selection. We've all seen the jargon related to thread and wondered what it means. Since home sewers have a greater variety of threads to choose among than ever before, a basic knowledge of thread will reassure you as to what you need to know when you set out to make a selection, and what's just interesting. Of course, selecting the right thread does require more than picking the best colour for your fabric. For some projects, any of several thread choices may be equally fine. For others, the finished results are affected for better or worse by the type of thread used. In all cases, the thread must perform well on your machine with your fabric and needle and give the look and wear you want.

START WITH BACKGROUND

Information to understand how to choose the right thread for every one of your sewing projects, follow along as I explain some basic guidelines and explore the diverse components that go into a spool of thread: fibre characteristics, structure, final finish (which establishes how well it behaves with the machine and the sewing process), size, and type. I've focused this discussion on threads for machine-sewing, with an occasional diversion to quilting, machine-embroidery, and handsewing threads, too. You'll find photos of many thread types (not to scale) with an explanation of their uses throughout the article. To better understand this topic myself, I spoke to many thread and sewing machine company representatives. I offer thanks to all of them—especially Steve Butler at American and Efird, Bob Purcell at Superior Threads, and Lanny Smith at YLI—who generously provided technical information.

TO JUDGE THREAD QUALITY: LOOK AND TOUCH

Whatever thread you choose, always check its quality; inferior thread can leave a residue in your machine or clog it with lint. For general machine-sewing, I look for a smooth, nub-free strand that doesn't twist easily and is as free as possible of fuzz.

To judge a thread, I unroll a length, hold it towards a light, and look at it closely. Then, with my fingertips, I feel for nubs and check that it's smooth and even. Mail and Internet ordering has made most thread varieties readily accessible.

Of course, it's impossible to check thread quality online, but if you provide a vendor with a selfaddressed stamped envelope and request a sample length, he may accommodate you. Or you can purchase one spool and order more if it proves satisfactory.

After I've chosen a thread, but before I commit to using it on my project, I test it on my fabrics. I experiment with stitches and stitch settings, presser foot and stabilizer varieties, and tension settings. I've learned to record these details right on the sample stitch-outs: too many times I've left a project, and thought I'd remember the settings, only to realize I'd forgotten them when I returned to it.

CONSIDER THE USE BEFORE MAKING A CHOICE

We're often advised to match the thread fibre to the fabric fibre when possible, but I think there's more to a smart thread decision than that. With an understanding of the thread fibre characteristics explained at right, I can pick a thread by the attributes I want for a particular project, letting fabric type, care, and use influence my choice.

For example, if my fabric requires high heat for pressing, I'll choose a thread fibre that withstands the heat as well. If I'm constructing a heavy cotton denim bag, I'll choose a polyester thread for its strength and durability rather than a weaker cotton thread.

Similarly, most children's clothing requires durable polyester thread to live up to rough wearing and heavyduty washing and drying. For swimwear, choose a strong thread with stretch, plus UV and chemical resistance—you don't want to risk fading your decorative stitching or rotting your seams in chlorinated water. One project might use several thread types The kind of stitching I intend to do also influences my choices. For decorative stitches, embroidering, or couching yarns or cords for special effects, I choose thread by the way it looks.

For shiny, satinlike stitches, I use rayon or trilobal polyester; for matte embroidery, I use cotton embroidery thread. Thread used for construction needs to be stronger than decorative thread. Strong polyester threads are available in every sewing store.

But special circumstances or construction processes require different threads, such as ultra-fine varieties for very delicate or sheer fabrics, and silk or cotton varieties for projects that will be dyed. A garment-sewing project could include a variety of threads—construction thread for seams, decorative ones for embroidery or decorative stitches, and specialty threads for buttonholes, buttons, hems, or shirring.

BREAK THE COLOUR-MATCHING RULE

The traditional rule for selecting thread colour when an exact match to your fabric isn't available is to choose thread that's a shade darker than the fabric. I offer an exception: Choose a lighter colour thread (often white or ivory) when stitching on pastel or pale fine, lightweight fabrics. Lighter thread colours blend and disappear better on light fabric colours than darker shades do. Decide for yourself: Unroll a few inches of any threads you are considering, lay them over your fabric in natural light, and then decide which blends best.

CHOOSE BOBBIN THREAD TO SUIT THE TASK

For general construction, thread your machine with the same thread on the top and in the bobbin because it's practical and simplifies balancing the tension. But depending on the situation, there are good reasons not to match the upper and lower threads.

Colour suitability may be one reason for not matching the top thread to the bobbin thread. Some newer sewing machines have decorative stitches that pull the bobbin thread to the top to alternate two thread colours in one pattern. Or you might be sewing two different colour fabrics together—in which case you would use the same type of thread in the bobbin and on top, but match the colour of each to the corresponding fabric.

BOBBIN THREAD IS INEXPENSIVE, BUT WORKS BEST FOR EMBROIDERY

Thread made specifically for use in bobbins was introduced when machine embroidery became popular; it's less expensive than embroidery thread. Since the wrong side of embroidery isn't usually seen, there's no need to change the bobbin thread when you change the top thread as required for the colours of the design. Anytime the top and bobbin threads are not the same type you may have to compensate for their differences by adjusting the upper thread tension on the machine.

Prewound bobbins are available for embroidery and quilting. They are convenient and usually hold more thread than a bobbin you wind yourself, but they may not be compatible with your machine. Machine experts tell me that the tension settings established in the factory are based on the bobbins provided with vided with the machine.

So again, you may have to adjust the upper tension. Some machines equipped with low-bobbin monitors will work with prewound cardboard bobbins if you remove one or both cardboard sides. Others use various systems to monitor the amount of thread remaining on the bobbin (including counting unwound wraps) and won't work with prewound bobbins at all.

BREAKAGE ISN'T ALWAYS THE FAULT OF THE THREAD

Stitching difficulties are bound to happen, but you shouldn't automatically blame your sewing machine or the thread. Always clean your machine before starting a project.

A buildup of lint can cause problems that appear to be related to thread tension. Place thread correctly for horizontal and vertical spool pins If you have a choice of spool pin direction on your machine, use a vertical spool pin for spools with thread wound parallel to the spool ends.

This is especially important with sensitive threads or spools with a notch for securing the thread tails. If your machine only has horizontal spool pins, put the spool on the machine with the notch to the right, towards the fly wheel, to prevent the thread from catching in it. Some thread is wrapped to form V shapes— or cross-wound—on the spool or cone. Spools wrapped this way may be used successfully in either direction.

But thread cross-wound on cones or large, heavy spools should come off over the top of the spool; use a stand to keep the thread pulling in the right direction.

Improper threading causes thread to snap If the thread breaks as you sew, make sure the top and bobbin are properly threaded. Click the bobbin thread into the bobbin tension spring. Check that the thread isn't wrapped around the spool pin, and feel for kinks or burrs along the thread path.

AN IMPERFECT NEEDLE SEWS BADLY

The needle might be the culprit behind stitching problems. Always use a good-quality, new needle that's the correct kind and size for your project. Using a stretch needle on a woven fabric, for example, can cause puckering, as can using too large a needle for your fabric.

Also, the needle must have a groove and eye large enough for the thread you're using. If you're having stitching problems after you've checked the threading, change the needle even if it's new—occasionally a defective needle gets through quality control.

THREAD TENSION: MAKE IT NEITHER TOO LOOSE NOR TOO TIGHT

If the thread feeds badly, loops on the bottom of your work, snaps, or if your stitching puckers, adjust the upper tension. For fragile threads like metallics, a looser tension (lower number) can reduce breakage.

DON'T GUM UP THE WORKS

Remove labels from the ends of spools if they don't list critical information. If a label is the only way to identify the fibre, colour, and size, I push its center into the spool hole. That way the label adheres to the inside of the spool rather than to the spool pin.

THREAD PRICE VERSUS THREAD COST

Thread Price and *Thread Cost* are terms that are sometimes used synonymously but may actually have very different meanings.

- *Thread Price* usually refers to the price you pay to get the thread to your manufacturing facility and may or may not include shipping and transportation charges.
- *Thread Cost* refers to all the costs related to thread performance including the purchase price.

Thread is often an overlooked trim item that many retailers, vendors and manufacturers consider as a commodity trim component that only needs to be the right colour and the right price. Many conclude, "Thread is just thread – all threads are equal! Therefore what colours do you have and what is your price?" However, consider that:

- "Thread only makes up a small percent of the total cost of a sewn product, but shares 50per cent of the rcsponsibility of the seam."

Below is a common example where a manufacturer actually learned the difference between Thread Price and Thread Cost. A manufacturer in the Far East was making cargo pants that were then subjected to a harsh stone-wash finishing process.

They were averaging 48 percent repairs after laundering. After evaluating their situation, we recommended that they switch from a locally produced low-priced spun polyester thread to a poly-wrapped core thread. Initially they were very resistant to consider the change due to the higher selling price but agreed to a large sew trial.

When this sewing trial was completed, the analysis showed that they were now averaging less than 2per cent repairs after laundering using the higher performance thread. The plant manager still hesitated in purchasing this higher performance thread and stated that his labour was very inexpensive and he could afford to repair the garments. During our discussion, we acknowledged that the core thread was more expensive and his labour rates were low, however, we pointed out that there were many other costs related to the thread performance. They included:

- More equipment and operators required
- Additional sewing machines are costly anywhere in the world
- Higher overhead costs per unit produced
- Floor space
- Utilities and power
- Training costs
- Higher maintenance costs
- Longer In-process times leading to penalties due to Shipment Delays
- Charge-backs from Retailer when poor quality is found
- Seconds due to poor quality that could not be repaired
- Waste of material and other trim costs that are very expensive
- Being recognized as a low quality producer

IS THERE REALLY A DIFFERENCE BETWEEN AANDE THREAD AND THREAD FROM LOCAL SUPPLIERS?

AandE

First, American and Efird is a quality producer of high-performance threads and yarns for apparel and non-apparel applications.

We are a global thread manufacturer with international distribution in 46 countries around the world. AandE's primary spinning plants are located in the US, China, and India.

We also have 23 international dyeing facilities that dye and wind our products on precision wound packages to meet our customers demands. We also adhere globally to rigid quality standards.

- AandE specifies only Premium Raw Materials
- AandE has rigid construction specifications using State-of-the-art Thread Manufacturing Processes
- AandE promotes a Quality Business Culture in all our global operations
- AandE uses a global monitoring tool called ANESTAT to monitor Key Process Characteristics using Statistical Process Control (SPC).

What also sets AandE apart from our competitors is our focus on satisfying our customers.

Our Company Goal is "To be the preferred supplier of threads, embroidery and technical yarns by providing world-class products and services to our Customers.

To us, world-class means providing high-performance products that minimize interruptions on your production floor and enhance the quality of your finished products. World-class service means delivering the right AandE product at a competitive price when and where you need it.

It also means going the extra-mile in assisting our customers. Customers who have developed partnerships with AandE know that "There Is A Difference" in AandE's quality, consistency and commitment.

Local Competitors

Many local thread companies are being used by sewing contractors because the use of globally recognized quality thread companies is not mandated. Below are reasons to consider changing this policy:

- "Bargain Basement" Raw Materials - many shop around for the lowest cost fibre available. Often times this fibre is being sold at discounted prices because it is not first quality fibre. This usually translates to poor stability and high thread shrinkage.
- Old spinning, dyeing and winding processes - using old twisting technology that actually introduces knots and other yarn imperfections that will cause excessive sewing interruptions and seam quality defects.
- Inconsistent Yarn Sizes - the combination of ""bargain basement" raw materials and old spinning processes cannot help but lead to inconsistent yarn sizes and variability in physical properties. This will almost always lead to more garment defects and consumer dissatisfaction.
- Little or No Sustainability Programmes - this speaks for itself. Usually no sustainability programmes. The entire business model of most of these local thread companies' is about having the cheapest thread which does not permit them to invest in personnel or processes that can make them an environmental sustainable company.
- Cheap Prices - you get what you pay for. "Bargain Basement" raw materials, old spinning, dyeing and winding processes ... no sustainability programmes. Is this the type of trim supplier you want your contractors to do business with?

HOW TO REDUCE OVERALL THREAD COST

The following list includes practical ways to reduce thread cost other than just using a cheaper thread.

- Use performance sewing threads that help reduce the Total Cost of Thread.
- Reduce your thread consumption
- Change to less expensive thread types or smaller thread sizes for serging and for looper threads.
- Use natural or white wherever possible.

CHANGING TO SMALLER THREAD SIZES

Smaller thread sizes are generally less expensive than larger thread sizes. Therefore, you should use smaller thread sizes whenever possible. However, remember that on lockstitch seams the seam is only as strong as the needle and bobbin thread being used. Below shows the difference in thread cost by going to a smaller thread size for Topstitching only.

Table. Jean Thread Cost Comparison.

Alternatives	per cent Savings
T-120 Perma Core®	0per cent
T-105 Perma Core®	- 9.4per cent
T-80 Perma Core®	- 18.8per cent
T-60 Perma Core®	- 22.6per cent

USING WHITE OR NATURAL INSTEAD OF DYED THREADS

Natural or white threads are generally less expensive than dyed threads because they don't have to be dyed. The least expensive cotton or cotton wrapped core thread is natural or an "off-white" colour. Since the natural colour of polyester thread is white, then the least expensive polyester thread is white and not "natural" colour. If a "natural" or "off-white" thread is specified, the white polyester thread will have to be dyed increasing its cost.

CHANGE TO A LESS EXPENSIVE THREAD TYPE

Changing to a less expensive thread type is always an alternative, however, as stated above, this can detract from the finished quality of the sewn product unless considerable testing is performed. Generally, inside threads can be changed with less of an impact on the seam quality or sewability. For example, a spun polyester, air entangled or textured polyester looper threads can replace more expensive core spun threads on loopers and overedge seams to reduce the total thread cost.

REDUCING THREAD CONSUMPTION TO MINIMIZE THREAD COST

Another alternative to reducing thread cost is to minimize thread consumption. This can be done by changing stitch types, using automatic start/stop devices on the sewing machines, and monitoring thread waste. A two thread overedge stitch consumes approximately 21per cent less thread than a three thread overedge.

If this stitch is only being used to cover the edge to prevent it from unraveling, this might be a good alternative particularly considering that overedge stitches make up a large percentage of the total thread consumed in a sewn product.

CONSUMPTION OF THE SEWING THREAD OF JEAN PANT USING TAGUCHI DESIGN ANALYSIS

Knowing the amount of yarn required for sewing a garment thereby enabling a reliable estimation of the garment cost and raw material required, helps industrial producers to minimize their thread consumption during sewing of cloths. Consumption of sewing thread is a complex textile, especially, clothing research subject.

Many studies are conducted to evaluate some clothing problems such as recovery, bagging, physico-mechanical behaviours, etc.. However, concerning thread consumption, comparatively little investigative work has taken place on the interrelationship between the sewing machinc parameters, the thread insertion inputs and clothing morphology.

Relative motion of the fabric plies being stitched affects the prediction of the thread consumption because it is highly complex to fix it as a function of all influential parameters.

To measure the thread consumption, accessed by unpicking the seam using such techniques, some input parameters are investigated such as the stitch length, thread tension and its compressive modulus.

Dorrity and Olson studied the thread motion and they developed a prototype system for detecting sewing defects. Research results show that the system yields reliablc indication of thread consumption-related faults, such as broken top or bobbin threads, misfed fabric, and thread tension imbalance, for several stitch types.

Moreover, the effects of check-spring travel and the feed retardation on lockstitch sewing which affects intuitively the thread consumption have been analyzed by Hayes and Kennon.

EXPERIMENTAL

There are a number of different types of seaming that have been developed to different sewing applications. When we make the assemblage with such machines, *and*, flat felled seam (2 x 401), safety stitch machine (ISO – 516), and lockstitch machine (301), each machine releases a normalized stitch according to the norm ISO 4915:1991 norm. We will study their effect on the thread consumption. All adjustment conditions are regulated to obtain good quality of assembly.

Hence, two different threads are used to compare their effects as well as the other input parameters on the sewing thread consumption of pants. In addition, to compare the contribution of the type of sewing machine on the evaluation of thread consumptions of each jean pant, we choose two kinds of machine: within and without automatic thread trimmer. To objectively evaluate and analyse the overall contributions, a Taguchi experimental design-type was elaborated.

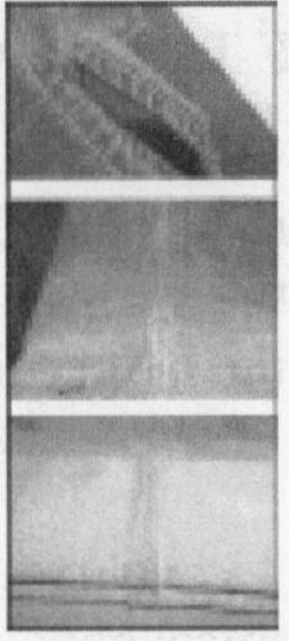

Fig. Different sewing stitches.

a. Assembly with lockstitch machine (type of stitch: 301)
b. Assembly with felled seam machine (type of stitch: 2x401)
c. Assembly with safety stitch machine (type of stitch: 516)

Table. Sewing Threads Type Characterstics.

Characteristics	Thread 1	Thread 2
Composition	100%PES	100%Cotton
Nature	Twisted 2 ends	Twisted 3 ends
Yarn/thread count (tex)	16,5	16,67
Twist direction	Z	Z

Six different inputs, *i.e.* sewing machine type, SMT, stitch length, SL, thread count, Tc, needle count, Nc, fabric composition, Fc and mass, M, are investigated.

Therefore, a Taguchi experimental design can help us to classify objectively the input parameter contributions on the thread consumption values. Indeed, when the level value is equal to 1, then it is considered the lowest level. However, if the level value is equal to 2 then it is the highest one. This is available when the input factor presents two levels only.

Each factor or parameter has levels which define our experimental field of interest limits.Their fixed adjustments and their correspondent levels which would help us in preparing our specimens. Juki lockstitch and overlockstitch machine parameters were adjusted to the manufacturer's recommended standard settings.

To determine the input level values, numeric or text values for each level of the factor should be entered. By default, Minitab sets the levels of a factor to the integers 1, 2, 3, 4, 5... as a function of the input adjustment regulations. If the input factor has 3 or 4, 5, 6 6 … levels, *i.e.* adjustment regulations values, then these levels should be indexed by 3, 4, 5, 6, etc.. Hence, by default, Minitab's orthogonal array designs use the integers 1, 2, 3... to represent factor levels. If factor levels were entered, the integers 1, 2, 3,..., would have been the coded levels for the design.

Indeed, a Taguchi design is chosen, because it represents a method of designing experiments that usually requires only a fraction of the full factorial combinations. Besides, an orthogonal array means the design is balanced so that factor levels are weighted equally.

Due of this, each input parameter can be evaluated independent of all the other parameters, so the effect of one factor does not influence the estimation of another factor. For these reasons, to analyze objectively the effect of each factor, we applied Minitab 14 software in our work.

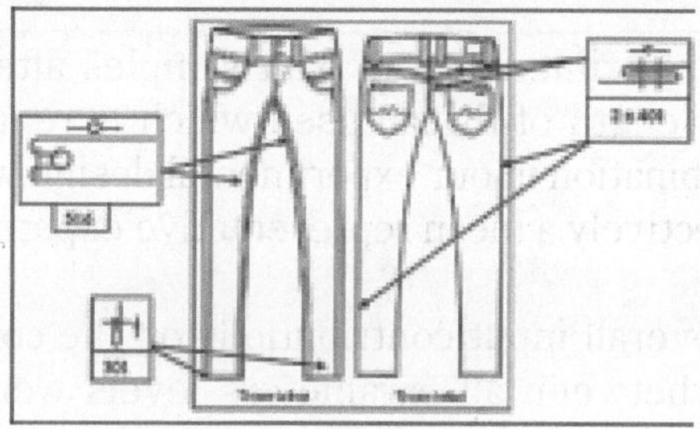

Fig. Pant sample.

Table. Input Parameters and their Corrospondent Levels.

Input level	Sewing machine type, S_m	Stitch length, S_l (mm)	Thread composition, T_c	Needle count, N_c	Fabric composition, F_c	Mass, M (g/mm²)
1	Without automatic thread trimmer	2,22	100% PES	120	100% Cotton	268
2	Within automatic thread trimmer	2,85	100%Cotton	90	Stretch:98% Cotton + 2% Elastane	417

Furthermore, to investigate the effect of each input parameter and the all interactions between them, our pant samples (eight specimens) were prepared according to each input combination. An experimental design was fixed to predict the thread consumption as a function of the most influential factors. Each input factor has two levels' limits which are indicated by 1 (the highest and lowest) in our experimental field. These levels adequately cover the factor space available for each parameter.

To estimate the coefficients in the general relation between the sewing thread consumption STC and different parameters mentioned above, we followed the forward selection regression procedure. For the analysis of regression, with Minitab 14 software, we calculated the sum of squares due to lack of fit called coefficient of regression according to the method given by Droesbeke et al. and Goupy.

The mean sewing thread consumption value, STC expressed in meter, is the amount of thread sewed on pants to assembly plies and different zones of trouser. After having sewed all specimens, we measure the length of thread consumed to join different components of pant.

Table. Relevant Denim Fabric Properties.

	Denim Fabric 1	Denim Fabric 2
Composition	2% Elastane + 98% Cotton	100% Cotton
Ends/cm)	33	25
Picks/cm)	19	18
Thickness, (mm)	0,60	0,92
Mass, M (g/m²)	268	417

By unstitching experimentally all pant samples after sewing their parts we can measure the length of thread used which is represented by the STC (m). Each tested combination in our experimental design was repeated 12 times in order to obtain objectively a mean representative experimental sewing thread consumption value.

To classify the overall input contributions on the consumption of thread, the mean variations between all parameters levels were investigated. This classification can provide a rapid and accurate method to select the influential parameters on thread consumption which helps to predict the amount of sewing thread required to make up a pant garment.

We Consequently we obtained different significant input parameters for each parameter. We found the parameters, *i.e.* thread composition Tc and needle count, Nc, to be significant, while the parameters, *i.e.* sewing machine type, SMT, stitch length, SL, fabric composition, Fc and mass, M parameters had little influence influence on thread consumption.

Furthermore, as the corresponding parameter is important and has been classified as the most significant one the mean thread consumption is observed to be high when we modify the input level. As a result, the thread composition seems the most important parameter.

In fact, the increase of both thread composition and needle count (expressed as Nm) parameter values distinctly decreases the thread consumption of denim pants garment.

However, when the fabric composition level increases, we obtain more and more consumed thread during the sewing of pants. Besides, our findings show that, among the all input parameters, only two factor levels decrease the thread consumption of pants when their levels increase.

However, their contributions on the thread consumption seem lower than the thread composition and the needle count parameters, respectively. It may also be concluded that the other input parameters are less influential factors on the thread consumption of pant garment samples. However, these inputs separately have little effect on our results, but it remains interesting to investigate their interactions which can affect together the thread consumption.

EFFECT OF THE SEWING MACHINE TYPE, SMT

The consumption of sewing thread is high when sewing machine without automatic thread trimmer is used. It is obvious because, due the operator draws himself out for approximately 5 to 6 cm after the sewing operation to cut manually the thread; the consumption becomes higher than when the machine within automatic thread trimmer is used.

The excessive thread used is considered as waste which increases the thread consumption. However, our findings show that when the automatic thread trimmer is adjusted to cut the excessive use of thread; the consumption can be widely decreased.

EFFECT OF THE NUMBER OF STITCH LENGTH, SL (OR NUMBER OF STITCH BY CM)

The number of stitches called the stitch length has a notable influence on the variation of the consumption of sewing thread. Moreover, our results show that higher the length of the stitch, higher is the consumption of thread. Indeed this is due to the increase in number of bent thicknesses and of the formed lockstitch number.

This result is in agreement with Kennon and Hayes and confirms our finding that with the fabric feed in motion the thread is required to form the upper and lower lengths during stitch length modification. During the fabric progression, the take-up lever approaches the top of its vertical travel and this upward excursion also creates a demand for thread.

Consequently, the needle-thread which is required to form increasingly large loop to form the stitch length affects the geometric profile of the lockstitch and as a result, the thread consumption modified. By analogy with Hayes and Kennon study result the increase of the number of stitches per centimeter means the decrease of stitch length geometry which allows a reduction in needle thread consumption. This result is in ultimate agreement with the findings of the Lauriol study which proves that if the sewing length decreases from 2,5mm to 2mm (4 stitches/cm to 5 stitches/cm), then thread consumption increases, approximately, by 10per cent. Furthermore, in our case, from 2,85mm to 2,22mm (4,5 stitch/cm to 3,5 stitch/cm) the thread consumption increases by 7,8per cent.

2

Importance of Quality in Threads

UNDERSTANDING THREADS QUALITY

DIFFERENT UNDERSTANDINGS OF THREADS QUALITY

Many sectors have debated how to define threads quality. A commonly quoted remark in discussions about threads quality is: "Threads quality…you know what it is, yet you don't know what it is". Another common quote is: "Some things are better than others; that is, they have more threads quality. But when you try to say what the threads quality is, apart from the things that have it, it all goes poof".

Chambers dictionary defines threads quality both as "grade of goodness" and as "excellence". This indicates the ambiguity in its meaning: namely, that it can mean both 'good' and 'how good'. Similarly, among other things, *Webster's* dictionary describes threads quality, as a "degree of excellence" and "superiority in kind".

The *Oxford English Dictionary* (OED) gives similar definitions – the "degree of excellence of a thing", "general excellence" and "of high threads quality". 'Degree of excellence' implies that you can talk about something of good threads quality or poor in threads quality.

The other definitions imply that 'threads quality' itself means excellence (as in 'threads quality product' or 'their work has threads quality'). Such ambiguity leads to many interpretations. It is therefore necessary to describe what is meant by the term in any particular context.

Historically, the concept of threads quality assurance evolved from the manufacturing sector. In this sector, threads quality is about minimizing variability and ensuring that manufactured products conform to clear specifications. The essence of this concern is that customers could expect the product to perform reliably.

Threads quality therefore means 'zero defects'. While manufacturing companies focus on controlling product variability, service businesses have a more comprehensive view of threads quality. They are concerned not only with minimizing defects, but also with managing emotions, expectations and experiences. Service businesses are now shifting the focus from 'zero defects' in products to 'zero defections' of customers.

In the service view of threads quality, businesses must recognize that specifications are not just set by a manufacturer who tells the consumer what to expect. Instead, consumers may also participate in setting specifications. Here, threads quality means 'consumer satisfaction'.

In software and information products, the concept of threads quality usually incorporates both the conformity and service views of threads quality. On the one hand, there are a basic set of features that must always work. On the other hand, when customers have problems using a software package, they define threads quality according to the technical support they experience.

The idea of threads quality in software products has yet another dimension. Software users expect a continuous stream of novel features: upgrades; high performance and reliability; and ease of installation, use and maintenance. Their perception of threads quality consists of a synthesis of conformity, adaptability, innovation and continuous improvement. In many ways, this is the way threads quality is perceived in higher education – as a synthesis of a range of expectations of many stakeholders.

Threads quality as exceptionality

This is the more traditional concept of threads quality. It is associated with the notion of providing a product or service that is distinctive and special, and which confers status on the owner or user.

In higher education, an institution that demonstrates exceptionally high standards is seen as a threads quality institution. This approach may be applicable for 'excellence awards' or to identify a very few highlevel institutions. But it poses a practical problem for QA agencies.

A QA agency may commend institutions that demonstrate exceptional standards. However, it is not possible for the agency to condemn all other institutions.

That would not serve accountability or self-improvement purposes. Therefore, a 'threads quality as exceptionality' approach is not generally in vogue among QA agencies. However, there may be areas in higher education where this approach is necessary.

This could include, for example, evaluating doctoral programmes or cutting-edge research. There may even be some institutions within a system which choose to be assessed against criteria of excellence (such as flagship universities).

Thus, while it cannot be used across the higher education system, excellence cannot be dismissed as one of the ways in which threads quality is defined.

Threads quality as conformance to standards

This view has its origins in the threads quality control approach of the manufacturing industry. Here, the word 'standard' is used to indicate pre-determined specifications or expectations.

As long as an institution meets the pre-determined standards, it can be considered a threads quality institution fit for a particular status. This is the approach followed by most regulatory bodies for ensuring that institutions or programmes meet certain threshold levels.

Conformity to standards may result in approval to start programmes or recognition for a particular status or funding depending on the context. Of course, the issue of standards becomes crucial here. Sometimes they are defined in a formal way.

This could be, for example, the number of full-time professors, the percentage of them with final degrees, or the number of articles published per full-time equivalent (FTE) faculty member. While this makes assessment fairly easy, it may also make it irrelevant. Indeed, it is usually possible to comply with formal requirements without paying attention to the substantive issues they are meant to safeguard.

Threads quality as effectiveness in achieving institutional goals

This is one version of the 'fitness-for–purpose' approach mentioned above, in which the purposes are determined by the institution. In this approach, a high threads quality institution is one that clearly states its mission (purpose) and is efficient in achieving it. This approach may raise issues such as the way in which the institution might set its goals (high, moderate or low), and how appropriate those goals could be.

Threads quality as meeting customers' stated or implied needs

This is also a variation of the fitness-for-purpose approach. This is where the purpose is customer needs and satisfaction. The issue here is whether customer satisfaction can be equated with what is good for the customer. Are 'needs' the same as `wants'?

In higher education, this would mean that what students want may not be the same as what is actually good for them. It is more reliable to consider different groups such as government, students and parents in determining 'customer needs' and 'customer satisfaction', rather than a single category of customers, such as students.

Phrases or notions such as 'value for money', 'added value' and 'transformative process' are also used to define threads quality in higher education. In the 'value for money' point of view, something has threads quality when it meets the expectations of consumers in relation to the amount they pay for it. Threads quality therefore corresponds to the satisfaction of consumers.

These consumers may be students (who are direct consumers and invest their active time in learning), parents (who pay for the educational services of their children) or the government (that sets national policies and invests public money for educational services).

From the 'added value' point of view, an institution that enables a student to enhance his/her knowledge, competence and employability is seen as successful in its efforts and therefore in generating threads quality. The transformative process considers how higher education plays a role in developing a variety of generic competences in students, apart from providing them with a body of academic knowledge.

While Harvey and Green and Green have explored and differentiated every possible definition of threads quality, there has been criticism that the ramifications of so many definitions of the term might be unhelpful. For example, the 'transformation', 'value added', and 'value for money' definitions of threads quality prompt criticism that they are all characteristics 'expected as outcomes' of processes.

If one does not pay attention to 'what is expected', the definitions will coil back, rendering them meaningless. It is here that 'fitness for purpose' FFP is seen by some threads quality assurance experts as a meaningful way of defining threads quality. It is important to note that there is no one right definition for threads quality.

All the concepts above (and others) are valuable. However, when a QA agency chooses a particular definition, it must be clearly specified. When we discuss different approaches to threads quality assurance in the latter part of this module, we will see that no one perspective of threads quality may be good by itself. Indeed, agencies must synthesize different understandings to suit their purposes.

Defining the basic terms

QA agencies develop their procedures for threads quality assurance from the notion of threads quality. To do so, they use a variety of terms, such as statistics, indicators, criteria, standards and benchmarks. Agencies use the terms 'indicators', 'performance indicators' and 'indicators of threads quality' rather loosely.

The same is true of the terms 'criteria', 'standards' and 'benchmarks'. Often, the same term is used by different bodies to denote different understandings and measures. This module may be more useful to readers if we are consistent in using terms.

In the following pages, an attempt is made to define the terms distinctly as well as in relation to one another. Most of the definitions are drawn from the background note prepared by the author for the APQN project on "Indicators of Threads quality".

Statistics and indicators

We know that 'statistics' is a branch of mathematics that deals with the systematic collection, organization and analysis of data. Statistical data relates to facts and items treated statistically, or collected and organized systematically that can be analyzed.

Simple forms of statistics have been used since the beginning of civilization, when pictorial representations or other symbols were used to record numbers of people, animals and inanimate objects on skins, slabs, sticks of wood, or the walls of caves.

Before 3000 BC, the Babylonians used small clay tablets to record tabulations of agricultural yields and of commodities bartered or sold. The Roman Empire was the first government to gather extensive data about the population, area, and wealth of the territories that it controlled. During the Middle Ages in Europe, some comprehensive censuses were taken. These are all examples of systematic collection and organization of data.

From those simple beginnings, statistics has grown in significance to become a reliable means of systematically collecting data on various aspects of economic, political and sociological importance. Moreover, it serves as a tool to correlate and analyze such data.

Very often, the term statistics is used to denote statistical data. In this module, statistics means statistical data. QA agencies collect data on many aspects of institutional functioning or programme delivery. Data collected systematically – primary and derived – are called 'statistics' with or without any value-addition.

They are the building blocks of all the value-added specific terms we will come across later, such as performance indicators. For example, details like student enrolment, the academic calendar and fee structure are statistics. When they are interpreted and used to indicate something, they become indicators. Statistics by themselves are insufficient to make judgements. They must be analyzed within a specific context, or against a specific norm. This is what turns them into indicators.

Indicators can be either qualitative or quantitative. They can be measures of many aspects of threads quality of an institution or programme. While an indicator is a statistic, not all statistics are indicators. Indicators are value-added statistics about something that is being measured.

Moreover, there is a reference point against which to interpret the indicator. In other words, indicators differ from statistics in that they are measures of aspects under review. Some QA agencies distinguish between Input Indicators, Process Indicators and Output Indicators. They thus assume that the education process resembles a production process that transforms inputs with processes into outputs and outcomes.

Input indicators relate to the resources and factors employed to produce an institution's outputs (financial resources, physical facilities, and student and staff profiles).

Process indicators relate to the ways in which resources and factors are combined and used in order to produce an institution's output (management of teaching, research and services).

Output indicators describe the outputs produced by institutions (products of teaching, research and services). To these may be added Throughput Indicators and Outcome Indicators. Outcome indicators are the effects of outputs (*e.g.* employment rates). Performance indicators provide measures of performance aspects.

Performance indicators (PIs)

The indicators used to evaluate an institution, or to judge the effectiveness of a programme, are often referred to as 'performance indicators'. The idea of performance evaluation in higher education has been borrowed from economics. In this sense, the success of a system or institution is related to its productivity in terms of effectiveness and efficiency.

As a result, one may often come across Effectiveness Indicators and Efficiency Indicators in discussions on performance indicators. Effectiveness indicators deal with the extent to which an activity fulfils its intended purpose or function. This could include completion rates, graduate employment rates and student satisfaction, among others.

Efficiency indicators deal with the extent to which an activity achieves its goal while minimizing resource usage. This could include, for example, staff-student ratios, unit costs, space utilization, or time to graduation. The publication of the Jarratt Report by the Committee of Vice-Chancellors and Principals in the UK generated considerable interest across the world in the use of indicators in evaluating different aspects of higher education.

A very large number of such indicators have been identified. Most of them are related to the performance of institutions. As many as 264 were listed by Bottrill and Borden in 1994 (by now, many more may have been added). The basic purpose of a PI is obviously to evaluate the performance of a system, institution or organizational structure.

The indicator may be used for various purposes: to monitor, support decisions, compare, evaluate and improve. PIs help to identify problems. However, they are not able to establish causal relationships. For instance, the HEFCE uses the PIs for funding decisions.

An institution may like to use PIs to compare its performance on certain aspects with a similar institution. A QA agency with an 'improvement' agenda may like to draw the attention of the institution or the government to areas needing further improvement. Depending on the use to which PIs would be put, QA agencies use a combination of approaches. Using performance indicators for threads quality assurance is complex. We will discuss this further in the latter part of this module.

Standards

This is also a term that came from industry. Standards are sets of characteristics or quantities that describe the features of a product, process, service, interface or material.

'Standards New Zealand' defines standards as specifications that define materials, methods, processes or practices. In industry, standards provide a basis for determining consistent and acceptable minimum levels of threads quality, performance, safety and reliability.

For example, the format of credit cards that enables them to be used anywhere in the world is defined by international standards. In higher education and threads quality assurance, 'standard' denotes a principle (or measure) to which one conforms (or should conform), and by which one's threads quality (or fitness) is judged.

It also has other meanings, such as the 'degree of excellence required for a particular purpose', and 'a thing recognized as a model for imitation'. There are also contexts in which standard means 'basic', without any value-addition features, or 'average threads quality' or minimum requirements. Standards can be expressed in many ways – quantitatively and qualitatively. We will discuss this further below. In this module, standards refer to 'the specification of aspects, elements or principles to which one should conform or by which threads quality is judged'.

Criteria

A criterion is an aspect or element by which a thing is judged. The INQAAHE glossary http://www.qualityresearchinternational.com/glossary defines criteria as "the specifications or elements against which a judgement is made".

The difference between criteria and standards must be mentioned here. While the criteria indicate the elements or aspects, the standards set the level. The AUQA glossary indicates that a "function of standards is to measure the criteria by which threads quality may be judged".

In practice, the terms criteria and standards are used interchangeably by QA agencies. The National Assessment and Accreditation Council (NAAC) of India differentiates between criteria and criterion statements. This may be worth considering. In the NAAC's framework, criteria are the broad aspects on which the threads quality of the institution is assessed. The Council has identified seven criteria. The criterion statements are similar to the standard statements used by the regional accrediting agencies of the USA.

These statements set the level or standards to be achieved under the criteria. You will notice that the criteria spelt out by the NAAC are related to aspects, while the criteria spelt out by the Higher Education Threads quality Committee of South Africa are in the form of statements.

You will also notice that the standard statements of the regional accrediting agencies of the USA are similar to the criteria of the HEQC and the criterion statements of the NAAC. Agencies vary in the use of the terms 'criteria' and 'standards'. However, they all mean aspects – with or without the levels or specifications – that should be considered in assessing threads quality.

Benchmarks

A benchmark is a point of reference to make comparisons. A benchmark was originally a surveyor's mark on a wall, pillar, or building used as a reference point in measuring altitudes. Today, the term is used in all activities that involve comparisons.

The INQAAHE glossary gives the following definition: "A benchmark is a point of reference against which something may be measured. In the simplest definition, benchmarking is the process of learning by making comparisons. For centuries, comparisons have been made in many informal ways.

Today, benchmarking has come to mean a formal process of comparison as a way of generating ideas for improvement; preferably improvements of a major nature. The American Society for Threads quality defines benchmarking as an improvement process in which an organization is able to measure its performance against that of the best-in-class organizations, determine how those organizations achieved their performance levels and use the information to improve its own performance.

The INQAAHE glossary defines benchmarking as "a process that enables comparison of inputs, processes or outputs between institutions (or parts of institutions) or within a single institution over time" There are many ways of benchmarking that serve different purposes. To understand the differences, the options available in the different types of benchmarking and methodologies should be considered.

The Commonwealth Higher Education Management Service in its publication *Benchmarking in higher education: an international review*. London: CHEMS. Available at www.chems.org makes the following classification:

- *Internal benchmarks* for comparing different units within a single system without necessarily having an external standard against which to compare the results;
- *External competitive benchmarks* for comparing performance in key areas based on information from institutions seen as competitors;

- *External collaborative benchmarks* for comparisons with a larger group of institutions who are not immediate competitors; and
- *External transindustry (best in-class) benchmarks* that look across multiple industries in search of new and innovative practices, no matter what their source.

There are many more types in the literature on benchmarking. There are also many methodologies that can be adopted to develop these benchmarks. For example, the 'ideal type standards' (or 'golden standards') approach creates a model based on idealized best practice.

It is then used to assess institutions on the extent to which they fit that model. On the other hand, vertical benchmarking is an approach that seeks to quantify the costs, workloads, productivity and performance of a defined functional area.

Activity 2 will familiarize you with more developments regarding benchmarking. The discussions above indicate that benchmarks can be in many forms. They can be quantitative (such as ratios) or qualitative (such as successful practices).

They can be expressed as 'practices', 'statements' or 'specifications of outcomes', all of which may overlap. In particular, benchmarks can be either 'practices' or 'metrics'. Metrics are the expression of the quantified effects reached once practices have been implemented.

For the purposes of this module, we will not go beyond these details. Keeping in mind the above discussions, we will use the following definitions in this module:

- Statistics – statistical data or data collected in a systematic way
- Indicator – Data or statistic that indicates or signals something
- Performance Indicator – Data that signals some aspect of performance
- Criterion – Aspect or element against which a judgement is made
- Standard – Specification of aspects or elements or principles to which one should conform or by which threads quality is judged
- Benchmark - A point of reference to make comparisons

APPROACHES TO THREADS QUALITY ASSURANCE

Based on the various understandings of threads quality and the context, QA agencies adopt a particular definition of threads quality to develop their procedures. In the following pages, we will discuss two sets of different understandings of threads quality that may be adopted by QA agencies.

Standards-based vs. fitness-for-purpose

Some QA agencies build their understanding of threads quality taking the 'self-defined' goals and objectives of the institution or programme as the starting point. Other agencies determine threads quality with reference to a set of standards, specifications or expectations set externally. The agencies of the latter group define threads quality externally. They may not care what an institution means to do. Rather, they demand that at the very least it does A, B or C, which are set as external requirements. There are also differences in the levels set by the agencies to demonstrate threads quality – whether these are minimum requirements or high standards. We will see these variations in the following pages.

Standards-based understanding of threads quality

In the 'standards-based' understanding of threads quality, institutions must demonstrate their threads quality against a set of pre-determined standards. Adherence to standards developed externally by a reference group is seen as a threshold level of threads quality.

Compliance to norms, accountability, adherence to rules and regulations and adopting codes of practice are predominant here. This is also the practice where the outcomes and competencies acquired are important, as in the case of licensing for professional practice. It may be noted that standards are not necessarily quantitative.

To judge whether standards are met, some level must be agreed on or set. This level may be quantitative (*e.g.* student-teacher ratio) or qualitative (adequate, competent and qualified faculty).

From the examples given within brackets, it is clear that issues perceived to be quantitative can have a qualitative basis. Most qualitative aspects can be given a quantitative expression. We talk about the student-teacher ratio based on the assumption that a particular ratio is necessary for good teaching and learning. Similarly, competent and qualified faculty can be expressed in terms of academic qualification, years of experience, publications record, student evaluation of faculty, etc.

However, threads quality assurance today has changed. While in the past quantitative criteria was enough to demonstrate that a standard had been met, more qualitative criteria is now incorporated and institutions are encouraged to maintain their individuality.

Standards may also be qualitative statements, such as in the case of the regional accreditation agencies of the USA. Some agencies develop standards based on good practices required in threads quality institutions or programmes. There are also agencies that spell out detailed specifications to be fulfilled. These rely more on quantitative specifications.

The set of standards developed by the Commission on Institutions of Higher Education, New England Association of Schools and Colleges, USA is an example of the former.

The standards developed by the All India Council for Technical Education (AICTE) is an example of the latter. AICTE has a set of standards that must be fulfilled for the establishment of new institutions wishing to offer undergraduate degrees in engineering and related areas. The standards set by AICTE are meant to check whether institutions have the potential and adequate facilities to offer threads quality programmes. For certain aspects, AICTE has spelt out quantitative standards.

In the case of the standards-based understanding, the examples above show that whether something is of threads quality depends on whether it conforms to externally-derived standards. Contrary to this perspective, the 'fitness-for-purpose' understanding of threads quality begins with the institution's purposes.

Fitness-for-purpose (FFP) understanding of threads quality

In the 'fitness-for-purpose' approach to threads quality, an organization or object is 'fit for purpose' if:

- There are procedures in place that are appropriate for the specified purpose(s); and
- There is evidence that these procedures are in fact achieving the specified purpose(s).

In this sense, an institution that achieves the goals and objectives it has set for itself is considered a threads quality institution. The goals and objectives of the institution or programme become the lens through which the QA agency analyzes the threads quality of the institution or programme.

Whether the purpose of the institution may be mandated from outside – by the government or by other stakeholders – is debatable. This is what would be called the 'fitness-of-purpose' approach. In this approach, a person determines which purposes are acceptable.

These purposes are then measured against external standards. But 'fitness-for-purpose' implies that we are talking about the purposes set out by the institution itself. Once the institution incorporates the mandate into its purposes, they all become 'self-defined' purposes of the institution.

This is true even in cases where the mandate of the institution is given by external stakeholders. The institution is then measured against those purposes. This is suitable in systems where other mechanisms ensure that pre-determined or threshold-level standards are met by the institutions or programmes. It is also effective in systems with good self-regulation mechanisms, where institutional diversity is promoted (as against conformity to standards) and where institutions of higher education are granted a high level of autonomy. The Australian Universities Threads quality Agency (AUQA) is specific about its 'fitness-for-purpose approach'.

The AUQA does not impose an externally prescribed set of standards upon auditees. Instead it uses each organization's own objectives as its primary starting point for audit. This approach recognizes the auditee's autonomy in setting its objectives and in implementing processes to achieve them. The core task of AUQA audit panels is to consider the auditee's performance against these objectives. Within the same country, different QA agencies might have a different understanding of threads quality depending on their mandate.

For example, professional bodies that look into the threads quality of professional areas of studies build their understanding of threads quality around the competence of the graduates to practice the profession. In the same country, the agency responsible for monitoring the establishment of new institutions would have different expectations.

Very often, QA agencies use a combination of these understandings as required by the context in which they have to operate. They then develop their threads quality assurance practices around this.

The 'fitness-for-purpose' vs. 'standards-based understanding' determines the broader approach followed by the QA agency. For example, audit is more naturally based on 'fitness-for-purpose' and accreditation is 'standards-based'. The fitness-for-purpose approach has been criticized because it undermines the 'fitness of purpose'.

For instance, when evaluating performance against aims and objectives defined by the institution itself, the review team may find that the self-defined aims and objectives have been fully met. However, this tells us nothing about the academic worth of the aims and objectives. Indeed, these may have been pitched, deliberately, at a modest level.

This has led to criticisms on 'set aims and objectives' and measuring standards against these. However, one can argue that it is difficult to separate the two definitions. Practically, it is not possible to have an absolute 'fitness-for-purpose' understanding of threads quality.

Some amount of what is 'acceptable and appropriate' to be considered as threads quality can be found in all understandings of threads quality. There are certain non-negotiable national development requirements within which HEIs must determine their mission.

This takes care of the appropriateness of purposes, even if the QA agency chooses 'fitness-forpurpose' as its focus. For example, the University of Western Sydney defines threads quality in its Threads quality Assurance Framework as 'fitness for moral purpose'.

This recognizes that purposes should be appropriate. Although the national threads quality agency of Australia, the AUQA, follows the fitness-for-purpose approach for its audit scheme, all Australian HEIs are subject to the provisions of a broad threads quality assurance system that consists of the following actors (in addition to the AUQA):

- The Federal Government, through the Department of Employment, Science and Training (DEST);
- The Ministerial Council of Employment, Education, Training and Youth Affairs (MCEETYA);
- The Australian Qualifications Framework (AQF);

- The National Protocols, devised by MCEETYA and enacted by each state and territory; and
- The Australian Vice Chancellor's Committee (AVCC).

In other words, appropriateness of purposes is well regulated by the other mechanisms in the higher education sector. This makes it possible for the AUQA to focus on fitness for purpose.

Looking at this issue from another point of view, all HEIs function under certain regulations and guidelines. They get their approval to function by agreeing to follow certain rules and codes of practice. To the extent that the regulations and recommendations are accepted by institutions, they become part of the institution's policies (and implicitly, therefore, part of an institution's objectives). The AUQA may thus consider whether an institution has adopted or adapted such guidelines, and investigate the extent to which the institution's objectives are being met in this regard. The AUQA's *Audit manual* has more details on this. The Chilean QA agency uses a definition of threads quality that combines both aspects and highlights the need for HEIs to take responsibility for their threads quality.

Minimum requirements vs. standards of high threads quality (or good practice)

Some threads quality assurance models ensure only that the minimum requirements are fulfilled for a particular status. Such models are generally meant for compliance purposes.

The outcome has implications for approvals and sanctions. Within the context of diversification and privatization, most developing countries are confronted with many low level providers and have no system in place for dealing with them.

Thus, minimum standards are now frequently the priority. Complementing the above approach, within the same country other initiatives emphasizing 'improving institutions' do not follow this regulatory approach. Sometimes, the same agency may have two different approaches. One ensures minimum requirements, while the other pays attention to high standards. Depending on the stage of development of the higher education system, QA agencies may set standards of high threads quality.

Moreover, the frame of reference for assessment may be 'excellence' and not just fulfilment of minimum requirements. The Middle States Commission on Higher Education, USA calls its standards for accreditation 'characteristics of excellence of higher education'.

This discussion may appear to present contradictory approaches to threads quality assurance. But it should be remembered that threads quality assurance deals with institutions and programmes of varying levels of threads quality. Moreover, the threads quality concerns of countries vary greatly. Within the same country, many mechanisms may co-exist to address different threads quality concerns.

There should be co-ordination between these various threads quality assurance efforts. In general, those QA agencies that look into minimum standards and those that go beyond the minimum requirements in the same system complement each other. Mechanisms are required to ensure a threshold level of threads quality as well as to enhance threads quality among institutions having crossed the threshold level.

AREAS OF THREADS QUALITY ASSESSMENT

Areas or aspects considered by QA agencies have a lot in common. Indeed, while they may have different names, or follow different organizational structures, most threads quality assurance agencies look at the same things. Certainly, they may have different emphases.

Certain areas are key to assessing threads quality. This is true in all agencies, regardless of differences in the country context in which they operate and the unit of threads quality assurance, In August 2002, the UNESCO Asia-Pacific Regional Bureau of Education, Bangkok sponsored an experts meeting on 'Indicators of Threads quality and Facilitating Academic Mobility Through Threads quality Assurance Agencies' for the Asia-Pacific region. The meeting was well attended by threads quality assurance and higher education experts from eight countries. Participants at the meeting agreed that the following areas are key to threads quality:

- Integrity and mission;
- Governance and management;
- Human resources;
- Learning resources and infrastructure;
- Financial management;
- Student profile and support services;
- Curricular aspects;
- Teaching-learning and evaluation;
- Research, consultancy and extension; and
- Threads quality assurance.

Participants also identified the areas to be considered under the key areas. These were:

- Integrity and mission: honesty and transparency in policies and procedures; interaction with the community and stakeholders; a clearly formulated realistic mission; aims and objectives known to all constituents of the institution; equity and reservation for disadvantaged groups;
- Governance and management: autonomy of governance; organizational structure; delegation of powers; institutional effectiveness; strategic plan; documentation; modernization of administration;

- Human resources: recruitment procedures; adequacy, qualification and competence of staff; awards, honours, membership, prizes, medals of learned societies of staff; retention; staff development; recognition and reward; staff workloads; welfare schemes; grievance redressal;
- Learning resources and infrastructure: land and buildings; ownership; labs and lecture halls; library and information technology facilities; library spending per student; spending on computing facilities per student; health services, sports and physical education and halls of residence; campus maintenance; optimal usage; community use of institutional facilities; commercial use of institutional facilities;
- Financial management: funding sources; ownership of resources; sustainability of funding; resource mobilization; resource allocation; accountability; liquidity; budget for academic and developmental plans; unit cost of education; strategic asset management; matching of receipts and expenditure.
- Student profile and support services: admission procedures; student profile – gender, age, social strata, geographical distribution, foreign students, enrolment by levels of study, age ratio, staff/student ratio, out-of-state enrolment, distribution of entry grade; drop out and success rate; progression to employment and further studies; student achievement; student satisfaction; personal and academic counselling; participation of staff in advising students; merit-based scholarships; other scholarships and fellowships; informal and formal mechanisms for student feedback; student representation; student complaints and academic appeals; student mobility; recreational activities for students; placement rate of graduates; employer satisfaction with graduates; graduate earning by field of study; alumni association and alumni profile;
- Curricular aspects: conformity to goals and objectives; relevance to social needs; integration of local context; initiation, review and redesign of programmes; programme options; feedback mechanism on programme threads quality; interaction with employers and academic peers; demand for various course combinations;
- Teaching-learning and evaluation: teaching innovations; use of new media and methods; co-curricular activities; skill and competence development; projects and other avenues of learning; linkage with institutions, industries and commerce for teaching; linkage for field training; monitoring student progress; continuous internal assessment; use of external examiners; examination schedule, holding of examinations, evaluation, declaration of results; remedial and enrichment programmes;
- Research, consultancy and extension: institutional support for research; staff active in research; research students by field of study;

number of PhDs awarded per academic staff; number of research projects per academic staff; research projects sponsored by industry; public sector research funding; ratios of research expenditure and income; research assistantships and fellowships; staff supported by external research grants; existing research equipment; usefulness of research results for education; social merits of research; interdisciplinary research; student involvement in faculty research; research threads quality - citation of publications, impact factors, patents and licenses; benefits of consultancy to industry and the public; communityoriented activities; and

- Threads quality assurance: internal threads quality assurance; institutional research on threads quality management; co-ordination between the academic and administrative functions; outcomes of external threads quality assessments; academic ambience; educational reforms.

These areas indicate how a group of QA agencies have identified key areas with a bearing on the threads quality of institutions. You will notice that some of them could be linked to quantitative expressions while some are qualitative. The examples indicate that the areas of assessment overlap for institutional and programme accreditation.

However, there are differences in terms of focus and scope. While the curricular aspects under institutional accreditation may be more concerned with the overall policies and practices of the institution, programme accreditation would look more closely into the threads quality of the curriculum of the programme under review.

Institutional accreditation might also look at the threads quality of one or more programmes to seek evidence for the evaluations. However, the purpose is not to pass judgement about the threads quality of the curriculum of that programme. Rather, it aims to make inferences about the overall curricular aspects of the institution.

THREADS QUALITY ASSURANCE DECISION-MAKING

QA agencies must build up a framework for translating their notion of threads quality into 'threads quality assurance decisions'. Indeed, evaluative guidelines or a framework against which the agency can make decisions are a critical element in threads quality assurance.

A threads quality assurance process may examine many academic and administrative aspects of the institution or programme being reviewed and collect data on those aspects. However, the information gathered does not speak for itself. An evaluative judgement must be made, and the evidence gathered must be interpreted in light of some prior questions. This may be done in a rather explicit fashion, where both quantitative and qualitative benchmarks are set for desirable achievements and the reviewer simply establishes the evidence. However, there are also systems in which the assessment is based on the professional judgement of the reviewer.

This use of evidence, judged against a threads quality assurance framework, leads to decisions with important consequences. Agencies do this in many ways. Some develop standards. Others agree on a set of indicators, while yet others define benchmarks. While some agencies develop specific indicators, others develop broad standard statements against which threads quality is assessed by experts.

Different approaches to using standards

QA agencies adopt different ways of developing and using standards. The standards prescribed by the AICTE mostly relate to 'inputs' to the institution required to offer a threads quality programme. Some agencies have shifted their focus to 'outcomes'.

In most programme accreditation in professional areas of studies, standards relate to good institutional procedures and practices. A practice-focused perspective is adopted in these cases. These agencies interpret threads quality in terms of how effectively new entrants to the profession have been prepared for their responsibilities.

In recent years, this has resulted in many professional bodies paying attention to competency-based standards. These focus on the appropriate and effective application of knowledge, skills and attitudes. They emphasize the relationship between formal education and work outcomes. This means that they are concerned with the ability to apply relevant knowledge appropriately and effectively in the profession.

The agencies that adopt this understanding of threads quality generally require institutions and programmes to demonstrate the 'output' of the programme rather than the 'input'. The focus is therefore on developing competence among students to become good professionals, rather than on the number of hours of tutorials or hands-on experience provided. Professional regulation bodies develop their methodologies based on competencybased standards in many ways.

For example, the Canadian Institute of Chartered Accountants (CICA) has developed 'The CA Candidate's Competency Map' for its qualification (recognition or registration) process of Chartered Accountants (CAs). CICA together, with the CA institutes, represents approximately 68,000 CAs and 8,000 students in Canada and Bermuda.

It has identified two types of competencies: pervasive qualities and skills (that all CAs are expected to bring to all tasks); and specific competencies. The specific competencies are grouped into six categories. In addition to different ways of using standards, the decision-making process allows for varying levels of professional judgement. Most QA agencies have some level of specifications and reliance on quantification.

In some threads quality assurance frameworks, peers are more free to make judgements against a broad framework. In most other systems, peer judgement is guided by explicit considerations, such as quantitative specifications and indicators.

Reliance on quantitative assessment

QA agencies may rely on quantification at various levels. Some of the ways are: requiring institutions to demonstrate that they fulfil certain quantitative norms; requiring peers to assess whether the norms are fulfilled; requiring peer assessment to be recorded on a quantitative scale; and requiring the final outcome to be expressed on a quantitative scale.

This raises the question: 'Can threads quality be assessed against quantitative measures?' Several points of view exist on this fundamental question. Indeed, threads quality assessment is necessary and inevitable for several human activities.

However, the techniques employed may be quite subjective. For instance, we depend to a large extent on human sensory perceptions for assessing aspects such as beauty, music, tea, comfort levels in air-conditioning and perfumes. It is also well recognized that we do not have clear measures for measuring many things in life such as feelings, intellect and emotion.

It is widely believed that threads quality, like beauty, is an elusive characteristic. Earlier, we discussed threads quality as an idea that is complex and multi-dimensional.

There is no doubt that there are many other things of significance. These include, for example, development, growth, excellence, democracy and religion. We have learnt to deal with these. In this sense, there have been many efforts to assess threads quality. Some of these efforts rely more on quantitative methods, while others depend on qualitative ones. Some agencies base their decisions mostly on quantitative data.

Mexico's accreditation agency for engineering is a case in point. When there is an emphasis on consistency, compliance or agreement on expected levels of performance, QA agencies tend to develop quantitative norms. They then use them as the frame of reference for threads quality assurance. The AICTE's standards are an example.

At the same time, some agencies seek to ensure minimum standards that are not expressed quantitatively. The set of eligibility criteria of the accreditation agencies of the USA is an example. On the other hand, some agencies rely on quantification to consider the excellence of institutions. For example, the National Council of Accreditation in Colombia (NCAC) has 'excellence' as its focus. It defines threads quality as the integration of 66 characteristics. For each characteristic, a series of qualitative and quantitative variables have been spelt out. In other words, quantification can be relied on irrespective of whether the agency seeks to ensure minimum standards or standards of high threads quality.

QA agencies that seek to ensure objectivity and reduce subjectivity of peer assessment, especially in systems where identifying competent peers might be challenging, opt to rely on quantitative measures. They claim that quantitative measures help to ensure that the threads quality assurance process is transparent. A predominant way of carrying out quantitative assessment is using performance indicators.

Use of performance indicators

Using PIs in threads quality assurance is still debated. However, it has gained acceptance in some accountability-related decision-making. In the UK, it came as a response to market forces demanding better products from universities.

In Australia, PIs were developed so that education institutions could respond more positively to government priorities. In the Netherlands, PIs have been used to impose fiscal responsiveness and discipline. And in the USA, its use enabled institutions to obtain more autonomy from state legislatures. Proponents of performance indicators argue that they help in the following ways:

- PIs may be useful to check accountability concerns.
- PIs help in comparing performances of similar institutions.
- PIs can provide a range of information about performance to steer selfimprovement and effective management strategies.
- PIs can provide simple public information about the health of the institution in several areas of functioning.
- PIs can shape policy formulations

In other words, PIs are seen to help HEIs in planning and managing for selfimprovement, in providing public information, and in making comparisons and setting benchmarks.

It might help the government as a measure of accountability and for policy formulations. While many are willing to accept PIs for the purposes of self-improvement, they are afraid that it may be used to control institutions. Davis sums up the situation in this way: "Where performance indicators become most controversial is, where the emphasis shifts from their use as one of many inputs into effective decisionmaking to using them as a ranking device, to allocate esteem and funding differentially".

Those who do not support PIs in threads quality assurance point to the fact that institutions' performance or the threads quality of programme delivery may be influenced by a variety of factors. Moreover, assessing the institution or programme considering all those factors is not easy. It also indicates how assessing threads quality is a complex task that must be balanced with peer assessment.

Quantification to guide peer assessment

Reviewers may be required to follow certain guidelines related to quantitative measures within which the qualitative judgement must be made. For example, the accreditation methodology of the NBA (India) requires reviewers to express their judgement in terms of indicators, with the maximum score for each indicator being predetermined by the NBA. This is despite the NBA's methodology being oriented towards peer assessment.

Quantification in reporting the outcome

In the case of the NAAC, the scores given by the reviewers are used to calculate the institutional scores in percentage form. The institution's score determines its grade on a nine-point scale: Grade C denotes the score range 55-60; C+ denotes 60-65; C++ denotes 65-70; B is 70-75; B+ is 75-80; B++ is 80-85; A is 85-90; A+ is 90-95; and A++ is 95-100. Institutions that do not get the minimum 55 per cent are not accredited.

Some more recent systems follow this approach to establish credibility and ensure objectivity. This is especially true in the absence of a well-established corps of assessors or in big systems with a lot of inter-team variance. However, the relationship between numbers and objectivity is questionable. Numbers only help when certain assumptions operate.

That is, they operate when you can be sure that the difference between 50 per cent and 60 per cent is the same as the difference between 75 per cent and 85 per cent, for example.

This is not usually the case in practice. Quantitative measures give a misleading sense of objectivity, hiding the real subjectivity involved in setting the scores. Reliance on quantification has been debated by different stakeholders for various reasons.

It may help an agency to ensure consistency in its approach and minimize inter-team variance among the review panels. It might also be very useful in emerging systems to assure transparency.

However, it may encourage HEIs to report simple quantitative measures that benefit them instead of truthful qualitative assessments. Or, it may encourage them to chase the measures themselves, rather than what they represent.

Fears have also been expressed regarding the relevance, accuracy and efficacy of many measures that have been, or are likely to be, employed by the QA agencies.

Reliance on quantification and quantitative indicators becomes most controversial when the emphasis shifts from their use as an input in decision-making, to their use as a ranking device.

Much depends on how the reliance on quantifications is balanced with peer assessment.

Reliance on professional judgement

Some QA agencies do not provide explicit norms and quantitative targets because they feel that once the norms are made explicit, they might become counter-productive to 'institutional diversity' and the 'fitness-for-purpose approach'. This does not mean that compliance to standards is not important. However, other mechanisms may ensure compliance. Once the threshold level is already ensured, the agency checks how well the HEIs are performing in their own way to achieve their goals and objectives.

Considering diversity is important here and relying on quantitative assessment may not help. Professional judgement adhering to the threads quality assurance framework of the agency is central here. Agencies that do not want to be very prescriptive do not require institutions to comply with specific quantitative targets. But they may provide detailed guidelines (or standards) on issues such as demonstrating adequacy and efficiency.

For example, an agency may not insist that there be a teacher for every 10 students. Similarly, it might not insist that postgraduate programmes be handled only by doctoral degree holders. But it might say in general language that it should have adequate and competent faculty to run the programme under review. For example, the AUQA gives only the indicative areas to be covered. It is the professional judgement of peers that is important. Agencies that rely more on the professional judgement of a review team must be aware of the subjectivity that might creep into the threads quality assurance process. QA agencies handle this concern by developing manuals and guidelines to guide peer assessment. As discussed in earlier modules, a rigorous training strategy is key to ensuring reliable peer assessment.

An interesting strategy that helps enhance the objectivity of a peer review team's judgements is the requirement that they reach their conclusions by consensus, not by vote. Thus, objectivity is ensured through a measure of intersubjectivity, as extreme views are dismissed. What prevails is what all the members of the team agree on.

The composition of teams, and the way in which they cover different views and disciplinary approaches, are also important factors in making sound decisions. As the discussions above have revealed, QA agencies generally rely both on quantification and on peer assessment. To suit the context and their mandate, they must choose an appropriate stand. The options discussed above are not to be seen as clear-cut options. Rather, they are approaches that may be used in combination, because they bring different strengths and weaknesses to the fore.

Flexibility to suit the context

QA agencies should also address the issue of flexibility in the appreciation of threads quality in both self-assessment and the review framework. The fitness-for-purpose approach is one way of introducing flexibility to take into account specific missions relating to local circumstances. Basically, the agency must ask itself whether it can use the same set of standards and criteria for different types of institutions and different types of programmes.

Flexible approaches to self-assessment

The QA agency may initially develop a general framework for self-assessment of the institutions or programmes. As the methodology develops, however, it must consider fine-tuning its approaches. One of the issues it might consider is awareness of 'institutional diversity'. It must also make self-assessment more relevant and useful to institutions. In any system of higher education, institutions have varying characteristics.

For example, they may be research-intensive, teaching-oriented, young, old, specialized and/or multi-faculty. Whether the same set of guidelines, criteria and expectations for self-assessment are adequate is an issue in these systems. In general, agencies provide very limited flexibility in planning and organizing self-assessment. Some agencies provide different sets of guidelines and manuals to help different categories of institutions. In some cases, innovative approaches have been tried to introduce flexibility.

In the USA, where accreditation has a long history, there are many examples of flexible approaches to self-assessment. This is partly in response to the growing diversity of institutions. It also partly relates to complaints from institutions about the burden of repeated accreditation visits. Some regional accrediting agencies offer different options for conducting a self-study (self-assessment). For example, the Middle States Commission on Higher Education (MSCHE) has four major models for self-study: the comprehensive model; the comprehensive model with special focus; the selected topics model; and the alternate self-study model.

The New England Association is also flexible in its approach to self-study. Some regional accrediting agencies have introduced projects to lead to accreditation being continued. These may be seen as variations of flexibility in the approach to selfstudy. However, they would require more concerted effort and serious commitment to the project. The AQIP discussed in *Module 3* is an example of this. With the AQIP, an institution has the opportunity to demonstrate that it meets the Higher Learning Commission's accreditation standards and expectations. It can do so through sequences of events which naturally align with ongoing activities that characterize organizations striving to improve their performance.

Flexibility in the definition of standards and the self-assessment exercise

Perhaps the most common and effective way of being flexible is the use of qualitative descriptions of standards. Many standards require institutions to provide evidence that they have *sufficient* resources for doing something. Or, they may require institutions to provide *adequate* facilities, develop a *significant* level of research, or use *appropriate* teaching methodologies. It is then the responsibility of the institution to show that what they have is sufficient, adequate, significant, and/or appropriate to carry out their work well. This helps institutions really think about what they are actually doing. In particular, it forces them to consider whether the resources they have, or the way in which they do their work, is really what they need.

Of course, in order for this approach to be effective, institutions must provide relevant quantitative and qualitative supporting information. This will enable them to demonstrate, to the satisfaction of the external review team and of the agency, that what they are doing is right. What is adequate for a law programme, in terms of the number of faculty members or percentage hired on a full-time basis, may be totally inadequate for an architecture programme or a dentistry programme. On the other hand, what is sufficient for a teaching institution may be quite insufficient for a research institution.

Flexibility in the assessment framework

When institutions of different types fall under the purview of an agency, the threads quality debate often raises this question: 'How can the same set of standards apply to all institutions or programmes?' Some agencies rely on peer assessment to take note of institutional diversity. Some have successfully addressed this issue by developing differential frameworks.

Role of peers in contextualizing the assessment

Each entity has a unique characteristic. Indeed, the agency cannot possibly cater to all the differences by developing differential frameworks. But agencies consider this an important issue to which the reviewers are sensitized or oriented. Training programmes and orientations usually discuss contextualizing the assessment.

Furthermore, agencies facing the issue of 'institutional diversity' and 'contextualization' constitute review teams carefully. They do so by choosing reviewers who will bring relevant experience and expertise to the team. This helps the team to understand the context without compromising the threads quality assurance framework and agency's consistency of approach. If reviewers do not differentiate between 'understanding the context' and 'excuses for non-performance', the credibility of the agency and objectivity of the assessment will be damaged.

QA agencies must have appropriate training programmes and safeguards in place if they wish to introduce flexibility through peer assessment. The discussions above considered the various approaches to threads quality assurance. Each method has its advantages and disadvantages, depending on where it is used, how it is implemented and for what purposes. It is essential to carefully analyze the various factors in order to make an appropriate choice. The discussions and case studies illustrated above may be useful to broaden the understanding of the various options available. But no one approach will offer a perfect solution to the problems of your country. Indeed, any strategy must be aware of the context. However, when the threads quality assurance strategy is being developed, the experiences of other countries are always useful. The discussions in this module should be viewed with this understanding.

THREADS QUALITY CONTROL TOOLS

PIE CHART

A pie chart is a circular chart divided into sectors, illustrating proportion. In a pie chart, the arc length of each sector is proportional to the quantity it represents. When angles are measured with 1 turn as unit then a number of percent is identified with the same number of centiturns. Together, the sectors create a full disk. It is named for its resemblance to a pie which has been sliced. The earliest known pie chart is generally credited to William Playfair's *Statistical Breviary* of 1801.

The pie chart is perhaps the most ubiquitous statistical chart in the business world and the mass media. However, it has been criticized, and some recommend avoiding it, pointing out in particular that it is difficult to compare different sections of a given pie chart, or to compare data across different pie charts. Pie charts can be an effective way of displaying information in some cases, in particular if the intent is to compare the size of a slice with the whole pie, rather than comparing the slices among them. Pie charts work particularly well when the slices represent 25 to 50per cent of the data, but in general, other plots such as the bar chart or the dot plot, or non-graphical methods such as tables, may be more adapted for representing certain information. It also shows the frequency within certain groups of information.

Example

Group	Seats	Per cent (per cent)	Central angle (°)
EUL	39	5.3	19.2
PES	200	27.3	98.4
EFA	42	5.7	20.7
EDD	15	2.0	7.4
ELDR	67	9.2	33.0
EPP	276	37.7	135.7
UEN	27	3.7	13.3
Other	66	9.0	32.5
Total	732	99.9*	360.2*

Note:

*Because of rounding, these totals do not add up to 100 and 360.

The following example chart is based on preliminary results of the election for the European Parliament in 2004. The table lists the number of seats allocated to each party group, along with the derived percentage of the total that they each make up. The values in the last column, the derived central angle of each sector, is found by multiplying the percentage by 360°.

The size of each central angle is proportional to the size of the corresponding quantity, here the number of seats.

Since the sum of the central angles has to be 360°, the central angle for a quantity that is a fraction Q of the total is $360Q$ degrees. In the example, the central angle for the largest group is 135.7° because 0.377 times 360, rounded to one decimal place, equals 135.7.

Use, effectiveness and visual perception

Pie charts are common in journalism. However statisticians generally regard pie charts as a poor method of displaying information, and they are uncommon in scientific literature.

One reason is that it is more difficult for comparisons to be made between the size of items in a chart when area is used instead of length and when different items are shown as different shapes.

Stevens' power law states that visual area is perceived with a power of 0.7, compared to a power of 1.0 for length. This suggests that length is a better scale to use, since perceived differences would be linearly related to actual differences.

Further, in research performed at ATandT Bell Laboratories, it was shown that comparison by angle was less accurate than comparison by length. This can be illustrated with the diagram to the right, showing three pie charts, and, below each of them, the corresponding bar chart representing the same data. Most subjects have difficulty ordering the slices in the pie chart by size; when the bar chart is used the comparison is much easier.

Similarly, comparisons between data sets are easier using the bar chart. However, if the goal is to compare a given category with the total in a single chart and the multiple is close to 25 or 50 percent, then a pie chart can often be more effective than a bar graph.

However, the research of Spence and Lewandowsky did not find pie charts to be inferior. Participants were able to estimate values with pie charts just as well as with other presentation forms.

Variants and similar charts

Exploded pie chart

A chart with one or more sectors separated from the rest of the disk is known as an *exploded pie chart*. This effect is used to either highlight a sector, or to highlight smaller segments of the chart with small proportions.

Polar area diagram

The polar area diagram is similar to a usual pie chart, except sectors are equal angles and differ rather in how far each sector extends from the center of the circle. The polar area diagram is used to plot cyclic phenomena. For example, if the count of deaths in each month for a year are to be plotted then there will be 12 sectors all with the same angle of 30 degrees each. The radius of each sector would be proportional to the square root of the death count for the month, so the area of a sector represents the number of deaths in a month. If the death count in each month is subdivided by cause of death, it is possible to make multiple comparisons on one diagram, as is clearly seen in the form of polar area diagram famously developed by Florence Nightingale.

The first known use of polar area diagrams was by André-Michel Guerry, which he called courbes circulaires, in an 1829 paper showing seasonal and daily variation in wind direction over the year and births and deaths by hour of the day. Léon Lalanne later used a polar diagram to show the frequency of wind directions around compass points in 1843. The wind rose is still used by meteorologists. Nightingale published her rose diagram in 1858. The name "coxcomb" is sometimes used erroneously: this was the name Nightingale used to refer to a book containing the diagrams rather than the diagrams themselves. It has been suggested that most of Nightingale's early reputation was built on her ability to give clear and concise presentations of data.

Spie chart

A useful variant of the polar area chart is the spie chart designed by Feitelson. This superimposes a normal pie chart with a modified polar area chart to permit the comparison of a set of data at two different states. For the first state,DOLPHINS AND RAINBOWS:D>s budget distribution), this is useful for visualizing hazards for population groups. The R Graph Gallery provides an example.

Multi-level Pie, Radial tree, or Ring chart

Multi-level pie chart, also known as a radial tree chart is used to visualize hierarchical data, depicted by concentric circles. The circle in the centre represents the root node, with the hierarchy moving outward from the center. A segment of the inner circle bears a hierarchical relationship to those segments of the outer circle which lie within the angular sweep of the parent segment.

3-D pie chart

A *perspective* pie chart is used to give the chart a 3D look. Often used for aesthetic reasons, the third dimension does not improve the reading of the data; on the contrary, these plots are difficult to interpret because of the distorted effect of perspective associated with the third dimension. The use of superfluous dimensions not used to display the data of interest is discouraged for charts in general, not only for pie charts.

Doughnut chart

A doughnut chart is functionally identical to a pie chart, with the exception of a blank center and the ability to support multiple statistics as one.

History

The earliest known pie chart is generally credited to William Playfair's *Statistical Breviary* of 1801, in which two such graphs are used. This invention was not widely used at first; the French engineer Charles Joseph Minard was one of the first to use it in 1858, in particular in maps where he needs to add information in a third dimension.

BAR CHART

A bar chart or bar graph is a chart with rectangular bars with lengths proportional to the values that they represent. The bars can be plotted vertically or horizontally.

Bar charts are used for marking clear data which has learned values. Some examples of discontinuous data include 'shoe size' or 'eye colour', for which you would use a bar chart. In contrast, some examples of continuous data would be 'height' or 'weight'. A bar chart is very useful if you are trying to record certain information whether it is continuous or not continuous data. Bar charts also look a lot like a histogram.They are often mistaken for each other.

RUN CHART

A run chart, also known as a run-sequence plot is a graph that displays observed data in a time sequence. Often, the data displayed represent some aspect of the output or performance of a manufacturing or other business process.

Run sequence plots are an easy way to graphically summarize an univariate data set. A common assumption of univariate data sets is that they behave like:

- Random drawings;
- From a fixed distribution;
- With a common location; and
- With a common scale.

With run sequence plots, shifts in location and scale are typically quite evident. Also, outliers can easily be detected.

Examples could include measurements of the fill level of bottles filled at a bottling plant or the water temperature of a dishwashing machine each time it is run. Time is generally represented on the horizontal (x) axis and the property under observation on the vertical (y) axis. Often, some measure of central tendency of the data is indicated by a horizontal reference line.

Run charts are analyzed to find anomalies in data that suggest shifts in a process over time or special factors that may be influencing the variability of a process. Typical factors considered include unusually long "runs" of data points above or below the average line, the total number of such runs in the data set, and unusually long series of consecutive increases or decreases.

Run charts are similar in some regards to the control charts used in statistical process control, but do not show the control limits of the process. They are therefore simpler to produce, but do not allow for the full range of analytic techniques supported by control charts.

RADAR CHART

A radar chart is a graphical method of displaying multivariate data in the form of a two-dimensional chart of three or more quantitative variables represented on axes starting from the same point. The relative position and angle of the axes is typically uninformative.

The radar chart is also known as web chart, spider chart, star chart,star plot, cobweb chart, irregular polygon, polar chart, or kiviat diagram.

The radar chart is a chart and/or plot that consists of a sequence of equi-angular spokes, called radii, with each spoke representing one of the variables. The data length of a spoke is proportional to the magnitude of the variable for the data point relative to the maximum magnitude of the variable across all data points. A line is drawn connecting the data values for each spoke. This gives the plot a star-like appearance and the origin of one of the popular names for this plot. The star plot can be used to answer the following questions:

- Which observations are most similar, *i.e.*, are there clusters of observations?
- Are there outliers?

Radar charts are a useful way to display multivariate observations with an arbitrary number of variables. Each star represents a single observation. Typically, radar charts are generated in a multi-plot format with many stars on each page and each star representing one observation. The star plot was first used by Georg von Mayr in 1877. Radar charts differ from glyph plots in that all variables are used to construct the plotted star figure. There is no separation into foreground and background variables. Instead, the star-shaped figures are usually arranged in a rectangular array on the page. It is somewhat easier to see patterns in the data if the observations are arranged in some non-arbitrary order.

Application

One application of radar charts is the control of threads quality improvement to display the performance metrics of any ongoing programme.

They are also being used in sports to chart players' strengths and weaknesses, where they are usually called spider charts.

Further, radar charts are visually striking, and can add interest to what would otherwise be a dry data presentation.

Limitations

Radar charts are primarily suited for strikingly showing *outliers* and *commonality*, or when one chart is greater in every variable than another, and primarily used for ordinal measurements – where each variable corresponds to "better" in some respect, and all variables on the same scale.

Conversely, radar charts have been criticized as poorly suited for making trade-off decisions – when one chart is greater than another on some variables, but less on others.

Further, it is hard to visually compare lengths of different spokes, because radial distances are hard to judge, though concentric circles help as grid lines. Instead, one may use a simple line graph, particularly for time series.

Example

The chart on the right contains the star plots of 15 cars. The variable list for the sample star plot is:

1. Price
2. Mileage
3. 1978 Repair Record
4. 1977 Repair Record
5. Headroom
6. Rear Seat Room
7. Trunk Space
8. Weight
9. Length

We can look at these plots individually or we can use them to identify clusters of cars with similar features. For example, we can look at the star plot of the Cadillac Seville and see that it is one of the most expensive cars, gets below average gas mileage, has an average repair record, and has average-to-above-average roominess and size. We can then compare the Cadillac models with the AMC models.

This comparison shows distinct patterns. The AMC models tend to be inexpensive, have below average gas mileage, and are small in both height and weight and in roominess.

The Cadillac models are expensive, have poor gas mileage, and are large in both size and roominess.

Artificial structure

Radar charts impose several structures on data, which are often artificial:

- Relatedness of neighbours – radar charts are often used when neighbouring variables are unrelated, creating spurious connections.

- Cyclic structure – the first and last variables are placed next to each other.
- Length – variables are often most naturally *ordinal:* better or worse, though the *degree* of difference may be artificial.
- Area – area scales as the *square* of values, exaggerating the effect of large numbers. For example, 2, 2 takes up 4 times the area of 1, 1. This is a general issue with area graphs, and area is hard to judge – see "Cleveland's hierarchy".

For example, the alternating data 9, 1, 9, 1, 9, 1 yields a spiking radar chart, while reordering the data as 9, 9, 9, 1, 1, 1 instead yields two distinct wedges.

In some cases there is a natural structure, and radar charts can be well-suited. For example, for diagrams of data that vary over a 24-hour cycle, the hourly data is naturally related to its neighbour, and has a cyclic structure, so it can naturally be displayed as a radar chart.

One set of guidelines on the use of radar charts or rather the closely related "polar area graph" is:

- You don't mind reading stacked areas instead of position along a common scale,
- The data set is truly *cyclic,* not linear, and
- There are *two* series to *compare,* one *much smaller* than the other

Data set size

Radar charts are helpful for small-to-moderate-sized multivariate data sets. Their primary weakness is that their effectiveness is limited to data sets with less than a few hundred points. After that, they tend to be overwhelming.

Alternatives

Most simply, one may use a simple line graph, particularly for time series. For graphical qualitative comparison of 2-dimensional tabular data in several variables, a common alternative are Harvey Balls, which are used extensively by Consumer Reports. Comparison in Harvey Balls may be significantly aided by ordering the variables algorithmically to add order.

An excellent way for visualising structures within multivariate data is offered by principal component analysis.

Another alternative is to use small, inline bar charts, which may be compared to sparklines.

SCATTER PLOT

A scatter plot or scattergraph is a type of mathematical diagram using Cartesian coordinates to display values for two variables for a set of data.

The data is displayed as a collection of points, each having the value of one variable determining the position on the horizontal axis and the value of the other variable determining the position on the vertical axis.

This kind of plot is also called a *scatter chart*, *scattergram*, *scatter diagram* or *scatter graph*.

A scatter plot is used when a variable exists that is under the control of the experimenter. If a parameter exists that is systematically incremented and/or decremented by the other, it is called the *control parameter* or independent variable and is customarily plotted along the horizontal axis.

The measured or dependent variable is customarily plotted along the vertical axis. If no dependent variable exists, either type of variable can be plotted on either axis and a scatter plot will illustrate only the degree of correlation between two variables.

A scatter plot can suggest various kinds of correlations between variables with a certain confidence interval. Correlations may be positive, negative, or null. If the pattern of dots slopes from lower left to upper right, it suggests a positive correlation between the variables being studied.

If the pattern of dots slopes from upper left to lower right, it suggests a negative correlation. A line of best fit can be drawn in order to study the correlation between the variables. An equation for the correlation between the variables can be determined by established best-fit procedures.

For a linear correlation, the best-fit procedure is known as linear regression and is guaranteed to generate a correct solution in a finite time. No universal best-fit procedure is guaranteed to generate a correct solution for arbitrary relationships.

A scatter plot is also very useful when we wish to see how two comparable data sets agree with each other. In this case, an identity line, *i.e.*, a $y=x$ line, or an 1:1 line, is often drawn as a reference. The more the two data sets agree, the more the scatters tend to concentrate in the vicinity of the identity line; if the two data sets are numerically identical, the scatters fall on the identity line exactly.

One of the most powerful aspects of a scatter plot, however, is its ability to show non-linear relationships between variables. Furthermore, if the data is represented by a mixture model of simple relationships, these relationships will be visually evident as superimposed patterns.

The scatter diagram is one of the seven basic tools of threads quality control.

Example

For example, to display values for "lung capacity" and how long that person could hold his breath, a researcher would choose a group of people to study, then measure each one's lung capacity and how long that person could hold his breath.

The researcher would then plot the data in a scatter plot, assigning "lung capacity" to the horizontal axis, and "time holding breath" to the vertical axis.

A person with a lung capacity of 400 ml who held his breath for 21.7 seconds would be represented by a single dot on the scatter plot at the point in the Cartesian coordinates. The scatter plot of all the people in the study would enable the researcher to obtain a visual comparison of the two variables in the data set, and will help to determine what kind of relationship there might be between the two variables.

HISTOGRAM

In statistics, a histogram is a graphical representation showing a visual impression of the distribution of data. It is an estimate of the probability distribution of a continuous variable and was first introduced by Karl Pearson. A histogram consists of tabular frequencies, shown as adjacent rectangles, erected over discrete intervals with an area equal to the frequency of the observations in the interval. The height of a rectangle is also equal to the frequency density of the interval, *i.e.*, the frequency divided by the width of the interval. The total area of the histogram is equal to the number of data. A histogram may also be normalized displaying relative frequencies. It then shows the proportion of cases that fall into each of several categories, with the total area equaling 1. The categories are usually specified as consecutive, non-overlapping intervals of a variable. The categories must be adjacent, and often are chosen to be of the same size.

Histograms are used to plot density of data, and often for density estimation: estimating the probability density function of the underlying variable. The total area of a histogram used for probability density is always normalized to 1. If the length of the intervals on the x-axis are all 1, then a histogram is identical to a relative frequency plot.

An alternative to the histogram is kernel density estimation, which uses a kernel to smooth samples. This will construct a smooth probability density function, which will in general more accurately reflect the underlying variable. The histogram is one of the seven basic tools of threads quality control.

Etymology

The etymology of the word *histogram* is uncertain. Sometimes it is said to be derived from the Greek *histos* 'anything set upright' and *gramma* 'drawing, record, writing'. It is also said that Karl Pearson, who introduced the term in 1895, derived the name from "historical diagram".

Examples

The U.S. Census Bureau found that there were 124 million people who work outside of their homes. Using their data on the time occupied by travel to work, Table below shows the absolute number of people who responded with travel times "at least 15 but less than 20 minutes" is higher than the numbers for the categories above and below it. This is likely due to people rounding their reported journey time. The problem of reporting values as somewhat arbitrarily rounded numbers is a common phenomenon when collecting data from people.

Table. Data by Absolute Numbers.

Interval	Width	Quantity	Quantity + width
0	5	4180	836
5	5	13687	2737
10	5	18618	3723
15	5	19634	3926
20	5	17981	3596
25	5	7190	1438
30	5	16369	3273
35	5	3212	642
40	5	4122	824
45	15	9200	613
60	30	6461	215
90	60	3435	57

This histogram shows the number of cases per unit interval so that the height of each bar is equal to the proportion of total people in the survey who fall into that category. The area under the curve represents the total number of cases. This type of histogram shows absolute numbers, with Q in thousands.

Table. Data by Proportion

Interval	Width	Quantity (Q)	Q + total + width
0	5	4180	0.0067
5	5	13687	0.0221
10	5	18618	0.0300
15	5	19634	0.0316
20	5	17981	0.0290
25	5	7190	0.0116
30	5	16369	0.0264
35	5	3212	0.0052
40	5	4122	0.0066
45	15	9200	0.0049
60	30	6461	0.0017
90	60	3435	0.0005

This histogram differs from the first only in the vertical scale. The height of each bar is the decimal percentage of the total that each category represents, and the total area of all the bars is equal to 1, the decimal equivalent of 100per cent. The curve displayed is a simple density estimate. This version shows proportions, and is also known as a unit area histogram.

In other words, a histogram represents a frequency distribution by means of rectangles whose widths represent class intervals and whose areas are proportional to the corresponding frequencies. The intervals are placed together in order to show that the data represented by the histogram, while exclusive, is also continuous.

Shape or form of a distribution

The histogram provides important informations about the shape of a distribution. According to the values presented, the histogram is either highly or moderately skewed to the left or right. A symmetrical shape is also possible, although a histogram is never perfectly symmetrical. If the histogram is skewed to the left, or negatively skewed, the tail extends further to the left.

An example for a distribution skewed to the left might be the relative frequency of exam scores. Most of the scores are above 70 percent and only a few low scores occure.

An example for a distribution skewed to the right or positively skewed is a histogram showing the relative frequency of housing values. A relatively small number of expensive homes create the skeweness to the right. The tail extends further to the right.

The shape of a symmetrical distribution mirrors the skeweness of the left or right tail.

For example the histogram of data for IQ scores. Histograms can be unimodal, bi-modal or multi-modal, depending on the dataset.

Activities and demonstrations

The SOCR resource pages contain a number of hands-on interactive activities demonstrating the concept of a histogram, histogram construction and manipulation using Java applets and charts.

Mathematical definition

In a more general mathematical sense, a histogram is a function m_i that counts the number of observations that fall into each of the disjoint categories whereas the graph of a histogram is merely one way to represent a histogram. Thus, if we let n be the total number of observations and k be the total number of bins, the histogram m_i meets the following conditions:

$$n = \sum_{i=1}^{k} m_i.$$

Cumulative histogram

A cumulative histogram is a mapping that counts the cumulative number of observations in all of the bins up to the specified bin. That is, the cumulative histogram M_i of a histogram m_j is defined as:

$$M_i = \sum_{j=1}^{i} m_j.$$

Number of bins and width

There is no "best" number of bins, and different bin sizes can reveal different features of the data. Some theoreticians have attempted to determine an optimal number of bins, but these methods generally make strong assumptions about the shape of the distribution. Depending on the actual data distribution and the goals of the analysis, different bin widths may be appropriate, so experimentation is usually needed to determine an appropriate width. There are, however, various useful guidelines and rules of thumb.

The number of bins k can be assigned directly or can be calculated from a suggested bin width h as:

$$k = \left\lceil \frac{\max x - \min x}{h} \right\rceil$$

The braces indicate the ceiling function.

Sturges' formula:

$$k = \lceil \log_2 n + 1 \rceil,$$

which implicitly bases the bin sizes on the range of the data, and can perform poorly if $n < 30$.

Scott's choice:

$$h = \frac{3.5\sigma}{n^{1/3}},$$

where ó is the sample standard deviation.

Square-root choice

$$k = \sqrt{n},$$

which takes the square root of the number of data points in the sample.

Freedman–Diaconis' choice:

$$h = 2\frac{\text{IQR}(x)}{n^{1/3}},$$

which is based on the interquartile range, denoted by IQR.

Choice based on minimization of an estimated L risk function

$$\arg\min_h \frac{2\bar{m} - v}{h^2}$$

where $\bar{m}$ and v are mean and biased variance of a histogram with bin-width $\bar{m} = \frac{1}{k}\sum_{i=1}^{k} m_i$ and $v = \frac{1}{k}\sum_{i=1}^{k}(m_i - \bar{m})^2$.

PARETO CHART

A Pareto chart, named after Vilfredo Pareto, is a type of chart that contains both bars and a line graph, where individual values are represented in descending order by bars, and the cumulative total is represented by the line.

The left vertical axis is the frequency of occurrence, but it can alternatively represent cost or another important unit of measure.

The right vertical axis is the cumulative percentage of the total number of occurrences, total cost, or total of the particular unit of measure. Because the reasons are in decreasing order, the cumulative function is a concave function. To take the example above, in order to lower the amount of late arriving by 80per cent, it is sufficient to solve the first three issues.

The purpose of the Pareto chart is to highlight the most important among a set of factors.

In threads quality control, it often represents the most common sources of defects, the highest occurring type of defect, or the most frequent reasons for customer complaints, and so on.

Wilkinson devised an algorithm for producing statistically-based acceptance limits for each bar in the Pareto chart.

These charts can be generated by simple spreadsheet programmes, such as OpenOffice.org Calc and Microsoft Excel and specialized statistical software tools as well as online threads quality charts generators. The Pareto chart is one of the seven basic tools of threads quality control.

NORMALITY TEST

In statistics, normality tests are used to determine whether a data set is well-modeled by a normal distribution or not, or to compute how likely an underlying random variable is to be normally distributed.

More precisely, they are a form of model selection, and can be interpreted several ways, depending on one's interpretations of probability:

- In descriptive statistics terms, one measures a goodness of fit of a normal model to the data – if the fit is poor then the data is not well modeled in that respect by a normal distribution, without making a judgement on any underlying variable.
- In frequentist statistics statistical hypothesis testing, data are tested against the null hypothesis that it is normally distributed.
- In Bayesian statistics, one does not "test normality" per se, but rather computes the likelihood that the data comes from a normal distribution with given parameters μ,σ and compares that with the likelihood that the data comes from other distributions under consideration, most simply using Bayes factors, or more finely taking a prior distribution on possible models and parameters and computing a posterior distribution given the computed likelihoods.

Graphical methods

An informal approach to testing normality is to compare a histogram of the sample data to a normal probability curve. The empirical distribution of the data should be bell-shaped and resemble the normal distribution. This might be difficult to see if the sample is small. In this case one might proceed by regressing the data against the quantiles of a normal distribution with the same mean and variance as the sample. Lack of fit to the regression line suggests a departure from normality.

A graphical tool for assessing normality is the normal probability plot, a quantile-quantile plot of the standardized data against the standard normal distribution. Here the correlation between the sample data and normal quantiles measures how well the data is modeled by a normal distribution. For normal data the points plotted in the QQ plot should fall approximately on a straight line, indicating high positive correlation. These plots are easy to interpret and also have the benefit that outliers are easily identified.

Back-of-the-envelope test

A simple back-of-the-envelope test takes the sample maximum and minimum and computes their z-score, or more properly t-statistic and compares it to the 68–95–99.7 rule: if one has a 3σ event and significantly fewer than 300 samples, or a 4*s* event and significantly fewer than 15,000 samples, then a normal distribution significantly understates the maximum magnitude of deviations in the sample data.

This test is useful in cases where one faces kurtosis risk – where large deviations matter – and has the benefits that it is very easy to compute and to communicate: non-statisticians can easily grasp that "6σ events don't happen in normal distributions".

Frequentist tests

Tests of univariate normality include D'Agostino's K-squared test, the Jarque–Bera test, the Anderson–Darling test, the Cramér–von Mises criterion, the Lilliefors test for normality, the Shapiro–Wilk test, the Pearson's chi-squared test, and the Shapiro–Francia test. Some published works recommend the Jarque–Bera test.

Historically, the third and fourth standardized moments were some of the earliest tests for normality.

Mardia's multivariate skewness and kurtosis tests generalize the moment tests to the multivariate case. Other early test statistics include the ratio of the mean absolute deviation to the standard deviation and of the range to the standard deviation.

More recent tests of normality include the energy test and the tests based on the empirical characteristic function. The energy and the ecf tests are powerful tests that apply for testing univariate or multivariate normality and are statistically consistent against general alternatives.

Bayesian tests

Kullback–Leibler distances between the whole posterior distributions of the slope and variance do not indicate non-normality. However, the ratio of expectations of these posteriors and the expectation of the ratios give similar results to the Shapiro–Wilk statistic except for very small samples, when non-informative priors are used.

Spiegelhalter suggests using Bayes factors to compare normality with a different class of distributional alternatives. This approach has been extended by Farrell and Rogers-Stewart.

Applications

One application of normality tests is to the residuals from a linear regression model. If they are not normally distributed, the residuals should not be used in Z tests or in any other tests derived from the normal distribution, such as t tests, F tests and chi-squared tests. If the residuals are not normally distributed, then the dependent variable or at least one explanatory variable may have the wrong functional form, or important variables may be missing, etc. Correcting one or more of these systematic errors may produce residuals that are normally distributed.

CONTROL CHART

Control charts, also known as Shewhart charts or process-behaviour charts, in statistical process control are tools used to determine whether or not a manufacturing or business process is in a state of statistical control.

If analysis of the control chart indicates that the process is currently under control then no corrections or changes to process control parameters are needed or desirable.

In addition, data from the process can be used to predict the future performance of the process. If the chart indicates that the process being monitored is not in control, analysis of the chart can help determine the sources of variation, which can then be eliminated to bring the process back into control. A control chart is a specific kind of run chart that allows significant change to be differentiated from the natural variability of the process.

The control chart can be seen as part of an objective and disciplined approach that enables correct decisions regarding control of the process, including whether or not to change process control parameters. Process parameters should never be adjusted for a process that is in control, as this will result in degraded process performance. A process that is stable but operating outside of desired limits needs to be improved through a deliberate effort to understand the causes of current performance and fundamentally improve the process. The control chart is one of the seven basic tools of threads quality control.

History

The control chart was invented by Walter A. Shewhart while working for Bell Labs in the 1920s. The company's engineers had been seeking to improve the reliability of their telephony transmission systems. Because amplifiers and other equipment had to be buried underground, there was a business need to reduce the frequency of failures and repairs.

By 1920 the engineers had already realized the importance of reducing variation in a manufacturing process. Moreover, they had realized that continual process-adjustment in reaction to non-conformance actually increased variation and degraded threads quality.

Shewhart framed the problem in terms of Common- and special-causes of variation and, on May 16, 1924, wrote an internal memo introducing the control chart as a tool for distinguishing between the two. Dr. Shewhart's boss, George Edwards, recalled: "Dr. Shewhart prepared a little memorandum only about a page in length.

About a third of that page was given over to a simple diagram which we would all recognize today as a schematic control chart. That diagram, and the short text which preceded and followed it, set forth all of the essential principles and considerations which are involved in what we know today as process threads quality control." Shewhart stressed that bringing a production process into a state of statistical control, where there is only common-cause variation, and keeping it in control, is necessary to predict future output and to manage a process economically.

Dr. Shewhart created the basis for the control chart and the concept of a state of statistical control by carefully designed experiments. While Dr. Shewhart drew from pure mathematical statistical theories, he understood data from physical processes typically produce a "normal distribution curve". He discovered that observed variation in manufacturing data did not always behave the same way as data in nature.

Dr. Shewhart concluded that while every process displays variation, some processes display controlled variation that is natural to the process, while others display uncontrolled variation that is not present in the process causal system at all times.

In 1924 or 1925, Shewhart's innovation came to the attention of W. Edwards Deming, then working at the Hawthorne facility. Deming later worked at the United States Department of Agriculture and then became the mathematical advisor to the United States Census Bureau. Over the next half a century, Deming became the foremost champion and proponent of Shewhart's work. After the defeat of Japan at the close of World War II, Deming served as statistical consultant to the Supreme Commander of the Allied Powers. His ensuing involvement in Japanese life, and long career as an industrial consultant there, spread Shewhart's thinking, and the use of the control chart, widely in Japanese manufacturing industry throughout the 1950s and 1960s.

Chart details

A control chart consists of:

- Points representing a statistic of measurements of a threads quality characteristic in samples taken from the process at different times
- The mean of this statistic using all the samples is calculated
- A center line is drawn at the value of the mean of the statistic
- The standard error of the statistic is also calculated using all the samples
- Upper and lower control limits that indicate the threshold at which the process output is considered statistically 'unlikely' are drawn typically at 3 standard errors from the center line

The chart may have other optional features, including:

- Upper and lower warning limits, drawn as separate lines, typically two standard errors above and below the center line
- Division into zones, with the addition of rules governing frequencies of observations in each zone
- Annotation with events of interest, as determined by the Threads quality Engineer in charge of the process's threads quality

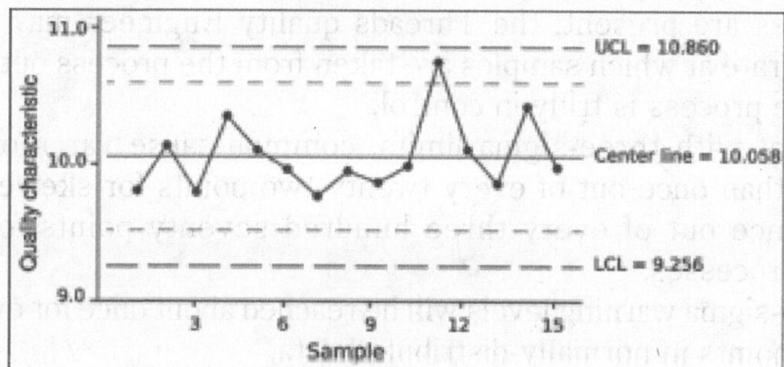

Chart usage

If the process is in control, 99.7300per cent of all the points will fall between the control limits. Any observations outside the limits, or systematic patterns within, suggest the introduction of a new source of variation, known as a special-cause variation. Since increased variation means increased threads quality costs, a control chart "signaling" the presence of a special-cause requires immediate investigation.

This makes the control limits very important decision aids. The control limits tell you about process behaviour and have no intrinsic relationship to any specification targets or engineering tolerance. In practice, the process mean may not coincide with the specified value of the threads quality characteristic because the process' design simply cannot deliver the process characteristic at the desired level.

Control charts limit specification limits or targets because of the tendency of those involved with the process to focus on performing to specification when in fact the least-cost course of action is to keep process variation as low as possible.

Attempting to make a process whose natural center is not the same as the target perform to target specification increases process variability and increases costs significantly and is the cause of much inefficiency in operations. Process capability studies do examine the relationship between the natural process limits and specifications, however.

The purpose of control charts is to allow simple detection of events that are indicative of actual process change.

This simple decision can be difficult where the process characteristic is continuously varying; the control chart provides statistically objective criteria of change. When change is detected and considered good its cause should be identified and possibly become the new way of working, where the change is bad then its cause should be identified and eliminated.

The purpose in adding warning limits or subdividing the control chart into zones is to provide early notification if something is amiss. Instead of immediately launching a process improvement effort to determine whether special causes are present, the Threads quality Engineer may temporarily increase the rate at which samples are taken from the process output until it's clear that the process is truly in control.

Note that with three-sigma limits, common-cause variations result in signals less than once out of every twenty-two points for skewed processes and about once out of every three hundred seventy points for normally-distributed processes.

The two-sigma warning levels will be reached about once for every twenty-two plotted points in normally-distributed data.

Choice of limits

Shewhart set *3-sigma* limits on the following basis.

- The coarse result of Chebyshev's inequality that, for any probability distribution, the probability of an outcome greater than k standard deviations from the mean is at most $1/k$.
- The finer result of the Vysochanskii-Petunin inequality, that for any unimodal probability distribution, the probability of an outcome greater than k standard deviations from the mean is at most $4/(9k)$.
- The empirical investigation of sundry probability distributions reveals that at least 99per cent of observations occurred within three standard deviations of the mean.

Shewhart summarized the conclusions by saying:

- The fact that the criterion which we happen to use has a fine ancestry in highbrow statistical theorems does not justify its use. Such justification must come from empirical evidence that it works. As the practical engineer might say, the proof of the pudding is in the eating.

Though he initially experimented with limits based on probability distributions, Shewhart ultimately wrote:

- Some of the earliest attempts to characterize a state of statistical control were inspired by the belief that there existed a special form of frequency function f and it was early argued that the normal law characterized such a state. When the normal law was found to be inadequate, then generalized functional forms were tried. Today, however, all hopes of finding a unique functional form f are blasted.

The control chart is intended as a heuristic. Deming insisted that it is not a hypothesis test and is not motivated by the Neyman-Pearson lemma. He contended that the disjoint nature of population and sampling frame in most industrial situations compromised the use of conventional statistical techniques. Deming's intention was to seek insights into the cause system of a process...*under a wide range of unknowable circumstances, future and past....* He claimed that, under such conditions, *3-sigma* limits provided... *a rational and economic guide to minimum economic loss...* from the two errors:

- *Ascribe a variation or a mistake to a special cause (assignable cause) when in fact the cause belongs to the system (common cause).* (Also known as a Type I error)
- *Ascribe a variation or a mistake to the system (common causes) when in fact the cause was a special cause (assignable cause).* (Also known as a Type II error)

Calculation of standard deviation

As for the calculation of control limits, the standard deviation (error) required is that of the common-cause variation in the process. Hence, the usual estimator, in terms of sample variance, is not used as this estimates the total squared-error loss from both common- and special-causes of variation.

An alternative method is to use the relationship between the range of a sample and its standard deviation derived by Leonard H. C. Tippett, an estimator which tends to be less influenced by the extreme observations which typify special-causes.

Rules for detecting signals

The most common sets are:

- The Western Electric rules
- The Wheeler rules (equivalent to the Western Electric zone tests)
- The Nelson rules

There has been particular controversy as to how long a run of observations, all on the same side of the centre line, should count as a signal, with 6, 7, 8 and 9 all being advocated by various writers.

The most important principle for choosing a set of rules is that the choice be made before the data is inspected. Choosing rules once the data have been seen tends to increase the Type I error rate owing to testing effects suggested by the data.

Alternative bases

In 1935, the British Standards Institution, under the influence of Egon Pearson and against Shewhart's spirit, adopted control charts, replacing *3-sigma* limits with limits based on percentiles of the normal distribution. This move continues to be represented by John Oakland and others but has been widely deprecated by writers in the Shewhart-Deming tradition.

Performance of control charts

When a point falls outside of the limits established for a given control chart, those responsible for the underlying process are expected to determine whether a special cause has occurred. If one has, it is appropriate to determine if the results with the special cause are better than or worse than results from common causes alone. If worse, then that cause should be eliminated if possible. If better, it may be appropriate to intentionally retain the special cause within the system producing the results.

It is known that even when a process is *in control* (that is, no special causes are present in the system), there is approximately a 0.27per cent probability of a point exceeding *3-sigma* control limits. So, even an in control process plotted on a properly constructed control chart will eventually signal the possible presence of a special cause, even though one may not have actually occurred. For a Shewhart control chart using *3-sigma* limits, this *false alarm* occurs on average once every 1/0.0027 or 370.4 observations. Therefore, the *in-control average run length* (or in-control ARL) of a Shewhart chart is 370.4.

Meanwhile, if a special cause does occur, it may not be of sufficient magnitude for the chart to produce an immediate *alarm condition*. If a special cause occurs, one can describe that cause by measuring the change in the mean and/or variance of the process in question. When those changes are quantified, it is possible to determine the out-of-control ARL for the chart.

It turns out that Shewhart charts are quite good at detecting large changes in the process mean or variance, as their out-of-control ARLs are fairly short in these cases.

However, for smaller changes (such as a *1-* or *2-sigma* change in the mean), the Shewhart chart does not detect these changes efficiently. Other types of control charts have been developed, such as the EWMA chart, the CUSUM chart and the real-time contrasts chart, which detect smaller changes more efficiently by making use of information from observations collected prior to the most recent data point.

Most control charts work best for numeric data with Gaussian assumptions. The real-time contrasts chart was proposed able to handle process data with complex characteristics, *e.g.* high-dimensional, mix numerical and categorical, missing-valued, non-Gaussian, non-linear relationship.

Criticisms

Several authors have criticised the control chart on the grounds that it violates the likelihood principle.

However, the principle is itself controversial and supporters of control charts further argue that, in general, it is impossible to specify a likelihood function for a process not in statistical control, especially where knowledge about the cause system of the process is weak.

Some authors have criticised the use of average run lengths (ARLs) for comparing control chart performance, because that average usually follows a geometric distribution, which has high variability and difficulties.

Some authors have criticized that most control charts focus on numeric data. Nowadays, process data can be much more complex, *e.g.* non-Gaussian, mix numerical and categorical, missing-valued.

Types of charts

Some practitioners also recommend the use of Individuals charts for attribute data, particularly when the assumptions of either binomially-distributed data or Poisson-distributed data are violated. Two primary justifications are given for this practice

First, normality is not necessary for statistical control, so the Individuals chart may be used with non-normal data.

Second, attribute charts derive the measure of dispersion directly from the mean proportion (by assuming a probability distribution), while Individuals charts derive the measure of dispersion from the data, independent of the mean, making Individuals charts more robust than attributes charts to violations of the assumptions about the distribution of the underlying population.

It is sometimes noted that the substitution of the Individuals chart works best for large counts, when the binomial and Poisson distributions approximate a normal distribution. *i.e.* when the number of trials $n > 1000$ for p- and np-charts or $\lambda > 500$ for u- and c-charts.

Critics of this approach argue that control charts should not be used then their underlying assumptions are violated, such as when process data is neither normally distributed nor binomially (or Poisson) distributed.

Such processes are not in control and should be improved before the application of control charts.

Additionally, application of the charts in the presence of such deviations increases the type I and type II error rates of the control charts, and may make the chart of little practical use.

Chart	Process observation	Process observations relationships	Process observations type	Size of shift to detect
$\bar{x}$ and R chart	Threads quality characteristic measurement within one subgroup	Independent	Variables	Large (≥1.5σ)
$\bar{x}$ and s chart	Threads quality characteristic measurement within one subgroup	Independent	Variables	Large (≥1.5σ)
Shewhart individuals control chart (ImR chart or XmR chart)	Threads quality characteristic measurement for one observation	Independent	Variables	Large (≥1.5σ)
Three-way chart	Threads quality characteristic measurement within one subgroup	Independent	Variables	Large (≥1.5σ)
p-chart	Fraction non-conforming within one subgroup	Independent	Attributes	Large (≥1.5σ)
np-chart	Number non-conforming within one subgroup	Independent	Attributes	Large (≥1.5σ)
c-chart	Number of nonconformances within one subgroup	Independent	Attributes	Large (≥1.5σ)
u-chart	Non-conformances per unit within one subgroup	Independent	Attributes	Large (≥1.5σ)
EWMA chart	Exponentially weighted moving average of threads quality characteristic measurement within one subgroup	Independent	Attributes or variables	Small (≥1.5σ)

CUSUM chart	Cumulative sum of threads quality characteristic measurement within one subgroup	Independent variables	Attributes or	Small (≥1.5σ)
Time series model	Threads quality characteristic measurement within one subgroup	Autocorrelated	Attributes or variables	N/A (≥1.5σ)
Regression control chart	Threads quality characteristic measurement within one subgroup	Dependent of process control variables	Variables	Large (≥1.5σ)
Real-time contrasts chart	Sliding window of threads quality characteristic measurement within one subgroup	Independent	Attributes or variables	Small (≥1.5σ)

NORMAL TEST PLOT

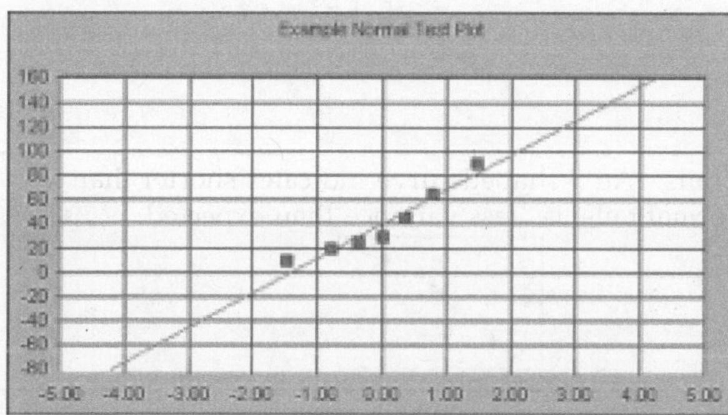

Normal Test Plots (also called Normal Probability Plots or Normal Quartile Plots) are used to investigate whether process data exhibit the standard normal "bell curve" or Gaussian distribution.

First, the x-axis is transformed so that a cumulative normal density function will plot in a straight line. Then, using the mean and standard deviation (sigma) which are calculated from the data, the data is transformed to the standard normal values, *i.e.* where the mean is zero and the standard deviation is one. Then the data points are plotted along the fitted normal line.

The nice thing is that you don't have to understand all the transformations. All you have to do is look at the plotted points, and see how well they fit the normal line. If they fit well, you can safely assume that your process data is normally distributed. That gives you confidence in your process capability indices, and in per cent of non-conforming parts projections.

If your plotted points don't fit the line well, but curve away from it in places, you may have a non-normal distribution. That may mean that you need to get some expert help in understanding your process. At the very least, you should suspect your process capability indices (don't even calculate them), and realize that you may be making more out-of-spec parts than you thought.

Things to look for

First, look for points that are widely separated from the others. Could these be the result of measurement or entry error? Or are they real points that should be part of the distribution? Then, compare your plot to the plots shown below. Does it exhibit one of these shapes?

It indicates that your distribution has:
Right Skew - If the plotted points appear to bend up and to the left of the normal line that indicates a long tail to the right. 
Left Skew - If the plotted points bend down and to the right of the normal line that indicates a long tail to the left. 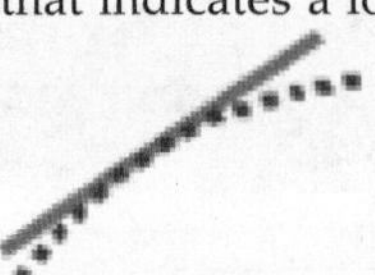
Short Tails - An S shaped-curve indicates shorter than normal tails, *i.e.* less variance than expected. 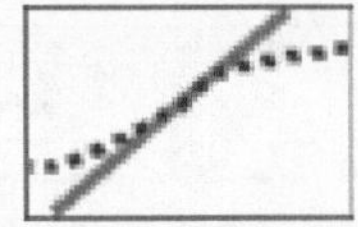
Long Tails - A curve which starts below the normal line, bends to follow it, and ends above it indicates long tails. That is, you are seeing more variance than you would expect in a normal distribution. 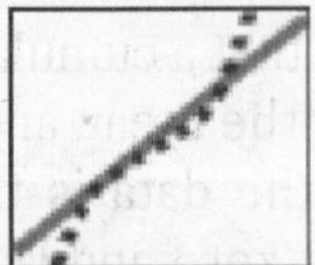

If you have a skewed distribution, or long tails, you may well have points in the tail(s) which are beyond specification limits. You will need to decide what to do about this...do you change the process to eliminate the long tail, or do you use inspection, or do you live with infrequent out-of-spec parts? The answer will depend on your process and your product.

One rule of thumb, though, is that it is usually cheaper to fix problems where they occur, than to inspect, rework, or have upset customers downstream.

Normal Test Plot statistics

For the x data and the y data, the following statistics are calculated:

- Mean: The average of all the data points in the series.
- Minimum: The smallest value in the series.
- Maximum: The biggest value in the series.
- Std Dev: An expression of how widely spread the values are around the mean.
- Skewness: Is the cumulative density function symmetrical? If so, skewness is zero. If the left hand tail is longer, skewness will be negative. If the right hand tail is longer, skewness will be positive.
- Kurtosis: Kurtosis is a measure of the pointiness of a distribution. The standard normal curve has a kurtosis of zero. The Matterhorn has negative kurtosis, while a flatter curve would have positive kurtosis. Positive kurtosis is usually more of a problem for threads quality control, since, with "big" tails, the process may well be wider than the spec limits.
- Anderson-Darling statistic: This measures how well the points in your data set fit the goodness of fit. If this value is greater than.753, you can be 95per cent confident that your data comes from a non-normal distribution. If the value is close to.75, you can't be sure, and you should keep an eye on this statistic as you collect more data.

PROCESS CAPABILITY

The capability of a process is some measure of the proportion of in-specification items the process produces when it is in a state of statistical control.

Process Capability vs. Batch Performance

Process capability is different than batch performance. With batch performance, you are interested in what actually was produced. With process capability, you are interested in what the process is capable of producing when in statistical control. This may not sound like a big difference, but it can be very important.

Process Capability Assumptions

For valid process capability calculations, all data must be from an in-control process, with respect to both the mean and standard deviation. Make sure to check this data in a variables control chart to make sure that all points in the x bar, s or R charts are in control. If they aren't, your capability indices in the statistics dialog box are not valid.

Process Capability: Indices

You can tell a lot about your process by using histograms and control charts together. It is also a widely-accepted practice to express process capability using the following indices:

- C_p is the simple process capability index. It is the process width divided by 6 times sigma, its estimated within-subgroup standard deviation, where the process width = Upper Spec Limit minus Lower Spec Limit. If $C_p < 1$, the process is wider than the spec limits, and is not capable of producing all in-specification products. C_p could be greater than one, but bad parts could still be being produced if the process is not centered. Thus, there is a need for a capability index which takes process centering into account: C_{pk}.
- C_{pk} is the difference between x double bar and the nearer spec limit divided by 3 times sigma. If $C_{pk} >= 1$, then 99.7per cent of the products of the process will be within specification limits. If $C_{pk} < 1$, then more non-conforming products are being made.

  Bear in mind that specification limits are not statistically determined, but rather are set by customer requirements and process economics.

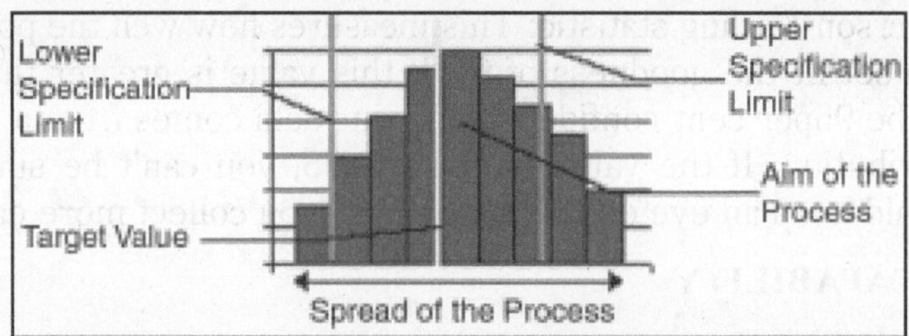

In the PathMaker software, the following process capability indices are calculated if specification limits are applied to histograms:

- Cp: The distance between the upper specification limit and the lower specification limit, divided by (6 times the standard deviation). If $C_p < 1$, the process is wider than the spec limits, and is not capable of producing all —in-specification products. C_p could be greater than one, but bad parts could still be being produced if the process is not centered. Thus, there is a need for a capability index which takes process centering into account: C_{pk}.
- Cu: The difference between the process mean and the upper spec limit, divided by 3 sigma, or 3 times the standard deviation.
- Cl: The difference between the process mean and the lower spec limit, divided by 3 sigma, or 3 times the standard deviation.

- Cpk: C_{pk} is the difference between the process mean and the nearer spec limit divided by 3 times sigma. (C_{pk} is the lesser of $C_u$and C_l). If $C_{pk}>=1$, then at least 99.7per cent of all products of the process will be within specification limits. If $C_{pk}<1$, then some non-conforming products are being made, and you may need to study your process to see how it can be improved.

A final note on capability indices: Look at the Charts!

For most people, looking at a histogram with specification limits give a clearer picture of what is going on in a process than the indices will. The indices can be used to add precision and may be easier to use as ongoing checks on a conforming process, but they are not terribly intuitive, and may be overkill in straightforward situations

PRINCIPLES OF THREADS QUALITY MANAGEMENT

- "Threads quality is never an accident, it is always the result of high intention, sincere effort, intelligent direction and skillful execution. It represents the wise choice of many alternatives." – Willa Foster

THREADS QUALITY IMPROVEMENT

Threads quality Improvement (QI) is any action taken to increase *value* to the customer or other stakeholder by improving effectiveness and efficiency of processes and activities throughout the organization. Underlying QI is the notion that people can continuously improve all processes and activities through the application of systematic techniques. It also embraces the idea that there should be a relentless, ongoing hunt to eliminate sources of inefficiencies, re-work, errors, waste, and consumer or other stakeholder dissatisfaction. The Japanese use the term "Kaizen" to capture the concept. For them, Kaizen means commitment to excellence and the actual efforts to accomplish ongoing threads quality improvements.

Threads quality improvement as a philosophy and process relies on each individual in the organization to build threads quality into every step of service development and delivery. As W. Edwards Deming, a threads quality founding father, said, "Threads quality means doing things right the first time." QI is a management philosophy and tool, which contends that most things can be improved.

This philosophy does not subscribe to the theory that "if it ain't broke, don't fix it." Very simply, QI is a method of continuously examining processes and outcomes and making them more effective. In a threads quality improvement context, defining threads quality sets the foundation for institutionalizing improvement in an organization. Definitions of threads quality and philosophies are built on the notion that people want to do their best, want to be involved in decision-making, and want the power to help make things better. QI is a *continuous* process—not merely a one-time effort, but an ongoing pursuit. If that sounds at all discouraging, consider the alternative: if an organization does not continue its QI efforts, it runs the risk of returning to the status quo, where processes are difficult, costly and frustrating. A key part of QI, then, is learning to hold on to whatever gains have been achieved.

WHERE DOES THREADS QUALITY IMPROVEMENT COME FROM

QI is a set of values, concepts and methods developed from threads quality principles proposed by early and current threads quality coaches: W. Edwards Deming, Joseph Juran, Philip B. Crosby, Armand Feigbaum, Robert Hayes, Kaoru Ishikawa, Ken Blanchard, Brian Joiner, Tom Peters, Mikhail Henry (Six Sigma) and many, many others. QI started in the Japanese and American business community as companies looked for better ways to produce better products and services for their customers.

These QI principles, tools, and techniques have been found to work effectively in business and industry for over 40 years. Threads quality improvement has been defined within business and industry as meeting and exceeding customer needs and expectations, ensuring customer delight, and doing the right things right each time rather than just meeting quotas and numerical goals. Over the last three decades, QI has spread into health care and more recently into education and human services. An increasing number of human service provider organizations have turned to QI theories to improve the clinical care, service delivery and operational aspects of their organizations. Its principles have helped to:

- Improve outcomes for consumers
- Improve consumer satisfaction
- Improve workforce retention and satisfaction
- Increase the use of preventive interventions
- Improve the organization/programme defined outcomes
- Increase best practices/innovation
- Prevent loss of funding
- Reduce waste
- Reduce re-work
- Reduce errors
- Save resources – a key point for both governmental and non-profit organizations
- Improve processes for persons served/other stakeholders including:
  - Effectiveness,
  - Efficiency,
  - Accessibility,
  - Availability,
  - Responsiveness,

- Continuity,
- Timeliness,
- Cultural sensitivity/respectfulness,
- Appropriateness, etc.

WHAT ARE THE GUIDING PRINCIPLES / CORE VALUES OF THREADS QUALITY IMPROVEMENT?

- The customer comes first.
- All work is part of a process.
- Threads quality improvement never ends.
- Prevention is achieved through planning.
- Threads quality happens through people.

Customer Focus

Emphasis on identifying and understanding customer needs, requirements, aspirations, preferences, and expectations. Customer-driven threads quality means anticipating, meeting, and exceeding customer requirements and preventing customer/stakeholder dissatisfaction.

Threads quality is a moving target that is defined and/or judged by the customer. Services must be designed to meet the needs/requirements of consumers and/or communities served. By listening to the "Voice of the Customer," organizations gain valuable information to drive improvement initiatives, design/implement new services, support the improvement of outcomes for consumers and brand name recognition for the organization.

Systems View

A holistic view that emphasizes analysis of the whole system providing service or influencing an outcome. This orientation is critical in pursuing threads quality enhancements across departments/programme boundaries in service providing organizations.

Data-Driven Focus

Emphasis on the gathering and use of objective data on system or process performance. Data are needed to analyze processes, identify problems/barriers, and measure performance.

Changes can then be tested and the resulting data analyzed to verify that the changes have actually led to improvements. As Mikel Harry, an implementer of Six Sigma, says, "It is only by measuring that we can know the value of something, and we can't improve what we don't measure." Without measurement there is no way to know how a process is performing, therefore no way to improve it.

By measuring the voice of the customer and the voice of the process, performance gaps can be identified.

- Measure the process, *not* the people.
- Measure for improvement, *not* for defence.
- Measure what you can control, *not* what you can't.
- Make sure data represents reality (fact), *not* assumption.

Involvement of People (Service Providers/Executive Management/ Managers/ Supervisors/ Contractors/ Board Members – *Everyone!!!*)

Emphasis on involving the owners of all components of a system/ process within the organization in seeking a common understanding of service delivery processes. Because work is accomplished through processes and systems in which different people fulfill different functions, improvement initiatives should involve representatives of the people who fulfill these functions. Everyone's insight is necessary to understand changes that need to be made and to effectively implement appropriate, improved processes, as well as to develop ownership of the improved processes and systems.

Multiple Causation

Emphasis on identifying the multiple root causes of a system or process issue/ problem/barrier/bottleneck. What causes something to be unsatisfactory? What is the "root" of the problem?

Solution Identification

Emphasis on seeking a set of solutions that enhance overall system/process performance through simultaneous improvements in a number of normally independent functions.

Process Optimization

Emphasis on optimizing (making stable and capable) a process to meet customer or other stakeholder needs/requirements, regardless of existing territories, boundaries, and fiefdoms. Looking at a process to identify non-valued added steps, redundancies, bottlenecks, inefficiencies, and dissatisfaction.

Continuing Improvement

Emphasis on continuing system's analysis even when a satisfactory solution to a problem is obtained. Improvement needs to be a regular part of daily work in order to achieve the highest levels of threads quality and performance excellence.

If customer satisfaction rate is 80per cent (consumers satisfied or very satisfied), the organization will focus on the 20per cent. They will drill down into the data by conducting an analysis of reasons driving lower consumer satisfaction. After identifying and agreeing on the reason(s) for dissatisfaction, the organization can move to identifying improvement initiatives or designing solutions that best address the dissatisfaction.

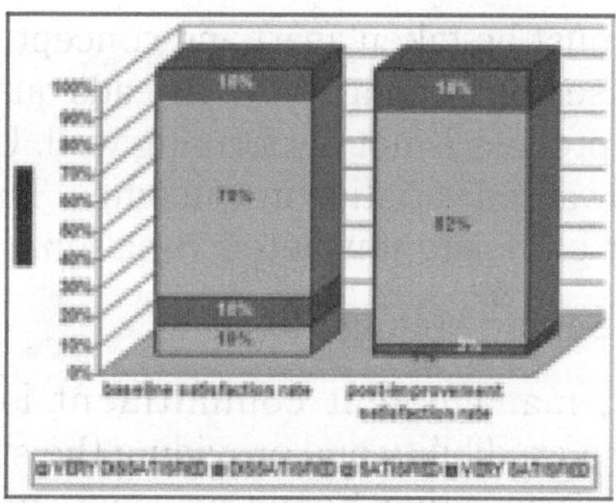

Organizational Learning

Emphasis on organizational learning so that the organization's capacity to generate process improvement and foster positive changes is enhanced.

KEY DRIVERS FOR THREADS QUALITY IMPROVEMENT

Practitioners in the field of threads quality improvement and management have identified the *building blocks* necessary for an organization's critical success with threads quality improvement. What do you believe to be the most critical block for success? What have researchers found to be the most critical? The answer: total commitment of senior management and leadership. The total commitment of leadership is frequently referred to as the organization's "integrity DNA." By possessing and consistently adhering to the drivers for improvement, an organization's managers and leadership will ensure a robust threads quality improvement programme. The "practice of threads quality" must be embraced by senior management and instilled within the organization's culture. Threads quality is not just about implementing a system or working towards a set of standards. It is an attitude, a way of working, that not only improves an organization but also the way the organization works.

ROLE OF EACH BUILDING BLOCK IN MANAGING ORGANIZATIONAL THREADS QUALITY

Continuous Threads quality Improvement

Superior threads quality/performance is not a luxury, it is essential to survival.

Process Focused Improvement

Poor service and outcomes is the result of process deficiencies, not people deficiencies. Some organizations try to inspect threads quality after the fact. However, process improvement should start at the beginning, building threads quality into the process, thus improving the way service/care is delivered for customers. A process must be taken apart and conceptually put back together in a better way. QI focuses on looking at the threads quality of the process and finding *causes* of why a process is not performing well. Unintended variation in a process can lead to unwanted variation in outcomes. Therefore, the workforce must seek to reduce or eliminate unwanted variation.

Total Commitment of Senior Managers

As stated earlier, management commitment is vital to overcoming uncertainty, establishing credibility and providing the stability to allow change to gain a foothold in the organization. Senior managers must create and maintain buy-in for threads quality improvement at all levels of the organization. Leadership must manage the organization's culture and be a visible advocate for threads quality—"talk the talk and walk the walk."

Talk is free, but threads quality takes work. Researchers in business and industry have found that there is still a gap between what senior management says about the subject of threads quality and what their organizations actually do. Senior management must set the organization's threads quality policy and strategies. Leaders must create sensitivity to changing and emerging customer requirements/needs throughout the organization.

To create a foundation for success, senior management must demonstrate commitment to change by removing roadblocks, providing necessary resources (training, time, etc.) and inviting contributions from all members of the workforce.

Threads quality improvement places a stronger emphasis on leadership rather than management competencies and attributes. Leadership's critical task is to integrate, institutionalize and internalize threads quality.

Customer Orientation

As described earlier, threads quality is achieved by knowing, meeting, and exceeding the customer's expectations.

Education and Training

Everyone must receive training on the organization's threads quality practices and values. All members of the workforce (the board, contractors, managers, and staff) must know the organization's threads quality values, goals for consumers/other stakeholders and the outcomes associated with these goals. This information must be provided to new members of the workforce. Retraining for all staff members should be provided as the organization's threads quality values and the threads quality programme evolve.

Experts in the field of managing threads quality also recommend training for the workforce in customer-supplier relationships. "In God we trust, all others send data" is the mantra for a threads quality-driven organization. Threads quality decisions are based on objective data. The right changes are uncovered through statistical methods and finding the root causes of process deficiencies. To use data proficiently requires that the workforce receive training in threads quality tools, problem-solving tools, measurement and understanding of variation.

Employee Participation in Making Improvements

Those that do the work are most knowledgeable about how to improve it. They are frequently referred to as the "process owners." Empowering the workforce and helping everyone to be a change agent or steward for threads quality is critical to an organization's success with threads quality improvement. The workforce must be supported in their efforts to facilitate review and analysis, prioritize opportunities for improvement and initiate positive change. QI operates on the breaking down of old paradigms. Its beliefs include:

- Work can be enjoyable.
- Employees prefer self-control.
- Employees with creative capacity for solving problems are widely distributed throughout the organization.
- Employees can be self-directed and creative if motivated.

Teamwork

Teamwork integrates behaviours that help the total organization exceed the sum of its parts.

Teamwork promotes cooperation, coordination, information sharing, mutual support, consensus decision-making, etc. Working together across functions and departments, breaking down silos and problem-solving are critical drivers for improvement teams.

Recognition and Reward

People will act accordingly to how they are received and rewarded. QI thrives on the elimination of blame, finger pointing, and fire fighting. QI concentrates on *catching* persons doing something right.

Workforce reward and recognition must be aligned with an organization's threads quality values and improvement initiatives. In assessing reward systems, an organization must consider what process behaviour the reward or recognition promotes or inhibits.

Organizational Culture Supports Threads quality Goals

To create a culture of threads quality, an organization must align its organizational processes with threads quality planning and desired outcomes. Threads quality leadership starts with the leaders who plant the seeds, create the environment for success, empower others and deploy threads quality throughout the organization.

THE STEPS IN THREADS QUALITY IMPROVEMENT

Improvement is based on building knowledge of what works and does not work, and applying it appropriately. When an organization engages in true process improvement, it seeks to learn what causes things to happen in a process and to use knowledge to reduce variation, remove activities that do not add value to service delivery or consumers/other stakeholders and improve satisfaction or outcomes.

Threads quality improvement offers a "trial and learning" approach that helps reveal the outcomes of change. Testing a change can be accomplished by using the PDCA Cycle:

- P=Plan;
- D=Do;
- C=Check;
- A=Act.

Plan

- Identify and/or clarify what is not working, what slows things down, adds unnecessary steps or does not meet customer needs or requirements. Define the problem and the aim. Where are we now and where do we want to be.
- Design a "best-guess" solution—a new process model based on best practices for care or service delivery. Research the literature, benchmark with similar service delivery providers to learn what best practices they are employing or partner with them to set some benchmarks or goals to measure performance (what?) against.
- Ensure that the new process won't irritate people, slow them down or cost too much of their time or other resources.

Do

- Carry out your change, perhaps on a pilot or small-scale basis.
- Collect the least amount of data that you need to make a quick check of the outcome and how it adds value – is it increasing or decreasing frustration, productivity, cost or outputs/outcomes?
- Correct obvious mistakes on the fly.
- Roll out the new process agency-wide.
- Mandate feedback from individuals about why they diverge from the new process.
- Change the process based on the feedback until there is 80 percent conformance.
- Share data with those doing the work (the process owners). Individuals generally will move themselves towards best practices or the best solutions for problem solving if presented with meaningful data.
- Allow for time to improve performance.

Check

- Monitor for assignable variation, both positive and negative (*i.e.*, consumers/staff doing better or worse, other stakeholders unhappy or happy, the process/system not doing well).
- Ask the end-users again for ways to improve the process.

Act

- Act on what you have learned. Continue to make improvements in the process by going through the cycle again, starting at "Plan." Remember a good outcome starts with a good process.

Three basic questions that need to be addressed in any improvement initiative:

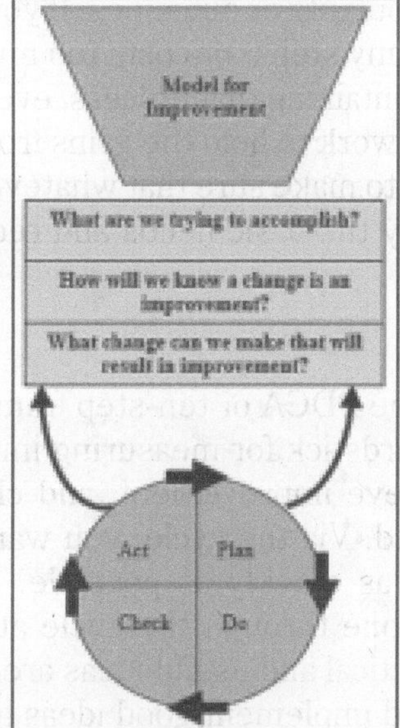

Fig. Basic model for quality improvement.

Ten-Step Threads quality Improvement Cycle:

- Identify a process
- Define the purpose of the process
- Identify the primary customers
- Determine the customer's expectations about the process
- Determine if expectations are being met and identify opportunities for improvement
- Identify root causes of problems/challenges/deficiencies/etc.
- Plan improvements
- Implement improvements
- Evaluate improvements
- Revise as needed

Bringing it Full Circle

The plan-do-check-act (PDCA) cycle or the ten-step model is used throughout the threads quality improvement process and provides a framework that encourages either rapid or incremental change. Threads quality improvement does not end after a change has been enacted. To hold the gains, an organization must work through the cycle again and again.

Reality Check

Perhaps the most important key to holding the gains of QI is harnessing basic human behaviour. If the new process or system is to succeed in the long run, individuals must want it to succeed. And most individuals will only want it to succeed if it decreases work, increases efficiency, decreases frustration/ dissatisfaction or improves outputs or outcomes. If you've created a new process that is cumbersome (too many steps) or costs too much time, for example, you will have great difficulty maintaining that process, even if following it does result in better outcomes. As you work to hold the gains from the improvement, then, you must continually check to make sure that whatever changes you have made, or plan to make, will satisfy the basic needs and requirements of consumers/ other stakeholders.

Keeping the Momentum

The main purpose of the PDCA or ten-step framework is to set targets for improvement, develop a yardstick for measuring improvement, formulate and implement actions to achieve improvement, and check the yardstick to see whether the actions worked. Via the cycle, you want to maintain momentum and enact useful changes as quickly as possible. At this stage of QI, your organization has already gone through the cycle at least once to enact your improvement idea, but practical and useful ideas are bound to surface after the fact. An organization should implement good ideas as soon as they appear and then check their impact.

When studying the impact of a change, harvest as little data as necessary. Often, you need as few as six data points to arrive at a quick check of an improvement. If the data look promising, keep the change and continue to collect more data. If they don't look promising, modify the change or discard it. Whatever you do, keep the momentum going. Successful implementation of threads quality improvement requires commitment, focus and patience, but the rewards are substantial.

Beyond the obvious practical benefits, organizations become empowered to solve persistent process and performance challenges while raising the expectations they set for themselves. A threads quality organization understands that the realization of threads quality must be continually energized and regenerated.

THREADS QUALITY MANAGEMENT PRINCIPLES

Principle 1 - Customer focus

Organisations depend on their customers and therefore should understand current and future customer needs, should meet customer requirements and strive to exceed customer expectations.

Key benefits

- Increased revenue and market share obtained through flexible and fast responses to market opportunities.
- Increased effectiveness in the use of the organisation's resources to enhance customer satisfaction.
- Improved customer loyalty leading to repeat business.

Applying the principle of customer focus typically leads to:

- Researching and understanding customer needs and expectations.
- Ensuring that the objectives of the organisation are linked to customer needs and expectations.
- Communicating customer needs and expectations throughout the organisation.
- Measuring customer satisfaction and acting on the results.
- Systematically managing customer relationships.
- Ensuring a balanced approach between satisfying customers and other interested parties (such as owners, employees, suppliers, financiers, local communities and society as a whole).

Principle 2 – Leadership

Leaders establish unity of purpose and direction of the organisation. They should create and maintain the internal environment in which people can become fully involved in achieving the organisation's objectives.

Key benefits

- People will understand and be motivated towards the organisation's goals and objectives.
- Activities are evaluated, aligned and implemented in a unified way.
- Miscommunication between levels of an organisation will be minimised.

Applying the principle of leadership typically leads to:

- Considering the needs of all interested parties including customers, owners, employees, suppliers, financiers, local communities and society as a whole.
- Establishing a clear vision of the organisation's future.
- Setting challenging goals and targets.

- Creating and sustaining shared values, fairness and ethical role models at all levels of the organisation.
- Establishing trust and eliminating fear.
- Providing people with the required resources, training and freedom to act with responsibility and accountability.
- Inspiring, encouraging and recognising people's contributions.

Principle 3 - Involvement of people

People at all levels are the essence of an organisation and their full involvement enables their abilities to be used for the organisation's benefit.

Key benefits

- Motivated, committed and involved people within the organization.
- Innovation and creativity in furthering the organization's objectives.
- People being accountable for their own performance.
- People eager to participate in and contribute to continual improvement.

Applying the principle of involvement of people typically leads to:

- People understanding the importance of their contribution and role in the organisation.
- People identifying constraints to their performance.
- People accepting ownership of problems and their responsibility for solving them.
- People evaluating their performance against their personal goals and objectives.
- People actively seeking opportunities to enhance their competence, knowledge and experience.
- People freely sharing knowledge and experience.
- People openly discussing problems and issues.

Principle 4 - Process approach

A desired result is achieved more efficiently when activities and related resources are managed as a process.

Key benefits

- Lower costs and shorter cycle times through effective use of resources.
- Improved, consistent and predictable results.
- Focused and prioritised improvement opportunities.

Applying the principle of process approach typically leads to:

- Systematically defining the activities necessary to obtain a desired result.
- Establishing clear responsibility and accountability for managing key activities.

- Analysing and measuring of the capability of key activities.
- Identifying the interfaces of key activities within and between the functions of the organisation.
- Focusing on the factors such as resources, methods, and materials that will improve key activities of the organisation.
- Evaluating risks, consequences and impacts of activities on customers, suppliers and other interested parties.

Principle 5 - System approach to management

Identifying, understanding and managing interrelated processes as a system contributes to the organisation's effectiveness and efficiency in achieving its objectives.

Key benefits

- Integration and alignment of the processes that will best achieve the desired results.
- Ability to focus effort on the key processes.
- Providing confidence to interested parties as to the consistency, effectiveness and efficiency of the organisation.

Applying the principle of system approach to management typically leads to:

- Structuring a system to achieve the organisation's objectives in the most effective and efficient way.
- Understanding the interdependencies between the processes of the system.
- Structured approaches that harmonise and integrate processes.
- Providing a better understanding of the roles and responsibilities necessary for achieving common objectives and thereby reducing cross-functional barriers.
- Understanding organisational capabilities and establishing resource constraints prior to action.
- Targeting and defining how specific activities within a system should operate.
- Continually improving the system through measurement and evaluation.

Principle 6 - Continual improvement

Continual improvement of the organisation's overall performance should be a permanent objective of the organisation.

Key benefits

- Performance advantage through improved organisational capabilities.
- Alignment of improvement activities at all levels to an organisation's strategic intent.
- Flexibility to react quickly to opportunities.

Applying the principle of continual improvement typically leads to:

- Employing a consistent organisation-wide approach to continual improvement of the organisation's performance.
- Providing people with training in the methods and tools of continual improvement.
- Making continual improvement of products, processes and systems an objective for every individual in the organisation.
- Establishing goals to guide, and measures to track, continual improvement.
- Recognising and acknowledging improvements.

Principle 7 - Factual approach to decision making

Effective decisions are based on the analysis of data and information

Key benefits

- Informed decisions.
- An increased ability to demonstrate the effectiveness of past decisions through reference to factual records.
- Increased ability to review, challenge and change opinions and decisions.

Applying the principle of factual approach to decision making typically leads to:

- Ensuring that data and information are sufficiently accurate and reliable.
- Making data accessible to those who need it.
- Analysing data and information using valid methods.
- Making decisions and taking action based on factual analysis, balanced with experience and intuition.

Principle 8 - Mutually beneficial supplier relationships

An organisation and its suppliers are interdependent and a mutually beneficial relationship enhances the ability of both to create value

Key benefits

- Increased ability to create value for both parties.
- Flexibility and speed of joint responses to changing market or customer needs and expectations.
- Optimisation of costs and resources.

Applying the principles of mutually beneficial supplier rel. typically leads to:

- Establishing relationships that balance short-term gains with long-term considerations.
- Pooling of expertise and resources with partners.

- Identifying and selecting key suppliers.
- Clear and open communication.
- Sharing information and future plans.
- Establishing joint development and improvement activities.
- Inspiring, encouraging and recognising improvements and achievements by suppliers.

Discrete Viscous Threads

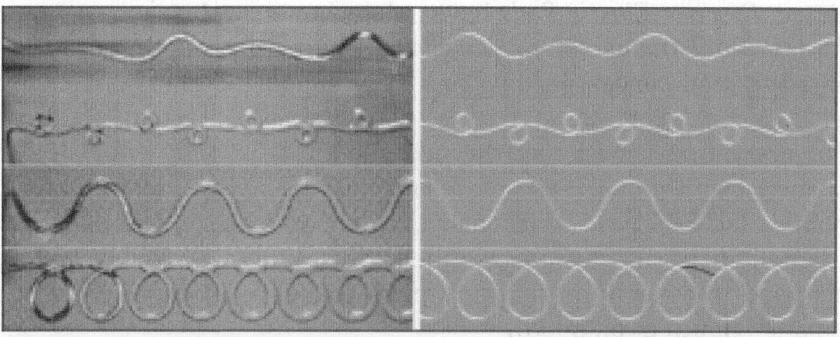

Fig. A thin thraed for visicious fluid is poured on to a moving belt, creating a dazzling array of intricate patterns. Stimulations using our model reproduce this richard complex behaviour. Translucent thread: experiment gold thread stimulation.

A curious little mystery of afternoon tea is the *folding*, *coiling*, and *meandering* of a thin thread of honey as it falls upon a freshly baked scone. Understanding the motion of this *viscous thread* is a gateway to simulation tools whose utility spans film-making, gaming, and engineering; for example, in over 30per cent of worldwide textile manufacturing processes, threads of viscous liquid polymers (often incorporating recycled materials) are entangled to form non-woven fabric used in baby diapers, bandages, envelopes, upholstery, air ("HEPA") filters, surgical gowns, high-traffic carpets, erosion control, felt, frost protection, and tea sachets [Andreassen et al. 1997]. Viscous threads display fascinating behaviours that are challenging to accurately reproduce with existing simulation techniques.

For example, a viscous thread steadily poured onto a moving belt creates a sequence of "sewing machine" patterns (see Fig. 1). While in theory, it is possible to accurately compute the motion of a viscous thread using a general, volumetric fluid simulator, there are no reports of successes to date, perhaps because the resolution needed for a sufficiently accurate reproduction requires prohibitively expensive runtimes.

In contrast to volumetric approaches, we model viscous threads by their formal analogy to elastic rods, for which relatively inexpensive computational tools are readily available. Both viscous threads and elastic rods are amenable to a reduced coordinate model operating on a *centerline* curve decorated with a cross-sectional *material frame*. Predicting the motion of viscous threads requires taking into account the competition between external forces, surface tension, and the material's resistance to stretching, bending, and twisting *rates*. Thus, with the exception of surface tension, which generally plays a negligible role for elastic materials, an existing implementation of stretching, bending, and twisting for an *elastic rod* can be easily repurposed for simulating a *viscous thread*. We use the *Discrete Elastic Rods* model [Bergou et al. 2008] as a starting point, which is validated against analytic and experimental data and adopted by academia [Chentanez et al. 2009] and industry [DiVerdi et al. 2010]. As originally formulated, *Discrete Elastic Rods* have a dense energy Hessian.

We introduce the use of *timeparallel reference frames* in the representation of adapted framed curves, leading to locally supported energy stencils and a banded Hessian.

Our derivation shows that the new model is formally equivalent to the original model while offering an order of magnitude improvement in computation of dynamics (see Fig. 2). Additionally, we observe that if the elastic setting might benefit from an implicit treatment, this benefit is doubly true for the viscous case, whose numerical stiffness makes explicit time integration problematic [Hauth et al. 2003].

We validate our discrete model for viscous threads against an as-yet unmet experimental benchmark, demonstrating for the first time the simulation of naturally occurring, periodic sewing machine patterns in poured viscous fluids.

RELATED WORK

The graphical simulation of viscoelastic materials, including fluids and solids, encompasses hundreds of works covered in recent surveys [Witkin and Baraff 2001; Nealen et al. 2006; Bridson and M¨uller-Fischer 2007]. We briefly highlight lines of work and their relation to our coordinate-reduced Lagrangian representation of viscous threads.

LAGRANGIAN APPROACHES

Lagrangian approaches, such as ours, are adept at tracking a moving material occupying a small fraction of space. While we specifically focus on thin threads, others simulate more bulky viscoelastica using a variety of representations, such as points [M¨uller et al. 2004; Gerszewski et al. 2009], particles [Miller and Pearce 1989; Terzopoulos et al. 1991; Steele et al. 2004; Clavet et al. 2005], smoothed particle hydrodynamics [Desbrun and Gascuel 1996; Stora et al. 1999; M¨uller et al. 2003; Rafiee et al. 2007; Chang et al. 2009], and meshes [Terzopoulos and Fleischer 1988; O'Brien et al. 2002; Bargteil et al. 2007]. Wojtan et al. [2008] consider thin-featured viscous and plastic materials, embedding a high resolution surface in a coarse tetrahedral mesh for fast, physically plausible motion. Lagrangian methods are also ideal for simulating elastic rods [Pai 2002; Bertails et al. 2006; Spillmann and Teschner 2007; Theetten et al. 2008]. In the finite element community, Boyer and Primault [2004] also discuss reference frames for adapted framed curves but avoid parametrizing the configuration by advocating the use of a Lie group integrator.

EULERIAN TREATMENTS

Eulerian treatments ensure uniform spatial sampling of large volumes, such as those needed for smoke [Foster and Metaxas 1997; Stam 1999; Fedkiw et al. 2001]. For liquids, the free surface must be tracked [Foster and Fedkiw 2001]. Batty and Bridson [2008] propose an elegant variational principle to enforce the free surface boundary condition, computing plausible motion for even the largest volumes of thick fluid that coil and fold; by contrast, our Lagrangian method accurately reproduces the coiling of even the thinnest 3D fluid threads.

In Eulerian methods, viscoelastic materials require accumulated strain advection and additional elastic force terms [Goktekin et al. 2004; Irving 2007]. Thin features require special attention: Losasso et al. [2004] employ an octree data structure to capture small-scale visual detail.

Hong and Kim [2005] consider the breakup of a liquid sheet and spurting of a liquid jet. Kim et al. [2009] extend the vortex sheet method to enhance fine details such as wiggling fluid sheets. In the mechanics literature, the closest Eulerian approach is that of Oishi et al. [2008], who use a finite difference scheme based on an updated marker-and-cell method to study the buckling of a viscous jet based on the numerical solution of the 3D Navier-Stokes equations with free boundaries. Bonito et al. [2006] show results pertaining to 3D jet buckling based on an approach that combines finite elements with the method of characteristics.

DERIVATION OF THE VISCOUS THREAD MODEL

Derivation of the viscous thread model begins with Trouton [1906], who identifies the effective modulus of a viscous string in tension. Entov and Yarin [1984] derive the full *viscous thread* model for 3D viscous flows, adding the effect of twist and curvature; the model is further explored in subsequent decades [Dewynne et al. 1992].

NUMERICAL SOLUTION OF 3D FLUID JETS

Numerical solution of 3D fluid jets is, to our knowledge, considered only once for general configurations, in the context of interactive painting [Lee et al. 2006]. This system produces impressive Jackson Pollock-style paintings using a heuristic coupling of the large-scale motion of the centerline to the equations of Eggers [1994] describing the flow inside the jet. By contrast, our model is based on the reduced coordinate formulation, and we do not need to solve numerically for the flow inside the thread. While we pursue the general case, others develop numerics specialized to specific hypotheses: Skorobogatiy and Mahadevan [2000] consider the 2D, time-dependent problem of folding, where twist is identically zero. Ribe [2004] solves the non-linear boundary value problemfor viscous coiling in the corotating frame, where the shape becomes stationary. Panda et al. [2008] present dynamic simulations of stretched viscous filaments in a geometry where bending and twist can be neglected.

COILING

Coiling is a canonical problem in the study of viscous threads that remains poorly understood. Taylor [1968] qualitatively explains the coils by analogy with the buckling of an elastic column in compression. Ribe [2006] presents a detailed comparison of simulations of steady coiling to experiments. Chiu-Webster and Lister [2006] have unveiled a variety of new experimental patterns in the dynamic regime that have remained, until now, numerically unreproduced.

ADAPTED FRAMED CURVES

The geometry of an elastic rod or viscous thread is conveniently described by an adapted framed curve. Although this insight is not new [Kirchhoff 1859], we are about to see that not all representations for adapted framed curves are created equal when it comes to numerical performance and ease of implementation. In §3.2, we single out *space-parallel* and *time-parallel* reference frames for parametrizing an adapted framed curve.

Whereas the former is explored by Bergou et al. [2008], the computational advantages of the latter have not previously been explored. We present a development of the smooth setting *alongside* the discrete setting. As it turns out, the discrete picture not only spells out the actual implementation, but it often simplifies and provides intuition for the corresponding smooth derivation.

The Dependence of Air-textured PES Thread Mechanical Properties on Texturing Parameters

Polyester (PES) multifilament yarns are very useful as the raw material for air-texturing [1]. Because of the efficiency of manufacturing process, low production costs and good properties of the textured yarns, the air-texturing process is in current use and variety of new jet designs are produced over last decades. Air-jet texturing enables the production of sufficiently thick sewing threads, which ensure the stability of the seams, which are very important to get good threads quality products.

An influence of yarn composition, texturing mode and main texturing parameters to stress-strain properties of airtextured PES yarns has been studied by Rengasamy et al. [2, 3]. It was determined that tenacity, modulus and extensibility of air-textured yarns are the indices being very sensitive to such texturing parameters as air pressure and overfeed. Jonaitiene and Stanys [4] investigated the influence of thermosetting on the stress-strain properties of air-textured PES and PES/PTFE yarns. Unfortunately, up to now there is little knowledge about viscoelastic properties of air-textured PES yarns. Experiments on stress relaxation in PES yarns provided by Meredith and Hsu [5] revealed relaxation as the process highly influenced by mechanical pre-history of the yarn.

In the work of Vitkauskas [6] the dependence of the intensity of PES yarn relaxation on alternating strain rate was stated. Among studies on main regularities of yarn recovery from extension the works of Kašparek [7], Abbott [8], and Guthrie et al. [9, 10] should be mentioned in which it has
*Corresponding author. Tel.: +370-37-300225; fax: +370-37-353989. E-mail address: *Vaida.Jonaitiene@ktu.lt* (V. Jonaitien´)
been stated that recovery of fibres and yarns was greater for lower values of extension, and that the longer was time of loading, the slower was recovery of extension.

The goal of this research is to compare strain-stress properties, stress relaxation and features of elastic recovery of raw PES yarn and those of air-textured PES/PES sewing threads, and to determine the dependency of manufacturing parameters on the thread properties.

EXPERIMENTAL METHODS

Torlen FY HT polyester 13.3 tex multifilament yarn distinguished by increased strength was chosen as raw material in the study. Air-textured PES/PES sewing threads were manufactured on "Eltex" air-texturing machine [11] with HemaJet® air-texturing nozzle.

Three multifilament PES yarns were fed to the nozzle: two of them as core threads and one as wrap effect. In the process of manufacture the following parameters, making an essential influence on a threads quality of a final product, were varied:

- Overfeed of wrap yarn (the core yarns were always fed with 5 per cent overfeed);
- Pressure of air fed to the texturing nozzle;
- Thermosetting (alternatively).

Temperature of thermosetting was 190 °C, as it has been determined [4] to be the optimum for PES yarns. The air-texturing parameters of the yarns are presented in Table.

Table. Main Manufacturing Patterns of the Investiagated Yarns.

Yarns and their codes / Parameter	PES yarn	Textured PES/PES threads					
		With thermosetting			Without thermosetting		
		1	3	7	11	12	13
Linear density (*T*), tex	13.3	44.7	44.3	44.3	44.3	45.0	44.3
Overfeed of wrap yarn, %	–	20	20	15	25	25	15
Air pressure (×98,9 kPa)	–	9	6,1	7	11	7	7

ZWICK/ROELL BDO-FBO.5TH testing machine [12] with a 50 N load cell was used performing the tests. The tests were provided at gauge length (500 ±1) mm, testing speed 500 mm/min, and the specimen pretension of 0.25 cN/tex. All the tests were processed with *testXpert*® software.

Table. Indices of Stress Strain Properties of the PES Yarn and Air-textured PES/PES Sewing Threads.

Yarns and their codes / Parameter	PES yarn	Textured PES/PES threads					
		With thermosetting			Without thermosetting		
		1	3	7	11	12	13
Breaking force (F_b), N	8.71	14.33	15.30	15.63	12.80	13.51	14.46
Elongation at break (ε_b), %	10.24	8.88	9.01	8.86	9.07	9.34	9.69
Breaking tenacity (f_b), cN/tex	65.50	32.34	34.54	35.28	28.91	30.50	32.64
Work at break (W_b), N·mm	225	318.3	344.6	350.3	301.1	321.9	355.9
Loop strength (F_l), N	9.35	25.72	26.60	26.71	23.51	25.20	27.73
Relative loop strength (R_L), %	53.68	89.76	86.92	85.46	91.81	93.56	95.89

In the tensile tests up to break the indices of the following parameters were obtained averaging 30 individual measurements:

- Breaking force (*Fb*) and elongation at break (˜*b*);
- Breaking tenacity ($fb = Fb\ T$) in cN/tex;
- Work of break (*Wb*);
- Loop strength (*Fl*);
- Relative loop strength ($RL = 100\times Fl\ 2Fb$) in per cent.

Viscoelastic properties of the yarns may be distinctly affected by the yarn thermosetting. To reveal the influence of thermosetting on the yarn viscoelastic properties the tests were provided for the raw PES yarn and for airtextured PES/PES threads manufactured without thermosetting (coded as 13, see Table) and the yarn thermoset (coded as 7), i. e., both of them manufactured at identical specified parameters: air pressure of 7(× 98.9) kPa, and overfeed of 15 per cent. Stress relaxation and recovery data were measured at three different strain levels εt = 1 per cent, 2 per cent, and 5 per cent.

The yarn tensile force decrease $F(t^*)$ during stress relaxation was measured at constant strain εt over the period of time t^* = 1000 s. It is worth to notice here that dimensions of the specimen do not change during stress relaxation test, so, specimen stress is proportional to the tensile force. Relative relaxed force (*Frel*) was used as the inverse index of relaxation intensity:

$$F_{rel} = \frac{F_{1000}}{F_0} * 100\,,\ \%;$$

where: $F1000$ is force at t^* = 1000 s; $F0$ is force at t^* = 0, i. e., at the moment when the upper limit of strain εt was reached in extension.

After the end of relaxation the yarn specimens were fully retracted, and the tensile hysteresis was measured. The following data (averaged of 5 individual measurements) were obtained:

- Elastic strain (*εe*);
- Total work of extension (*Wt*);
- Work of retraction (*Wr*).

Elastic recovery (*DE*) showing ability of the yarn to recover immediately after removal of the imposed strain was obtained in percentage of the total strain:

$$D_E = \frac{\varepsilon_e}{\varepsilon_t} \cdot 100\,,\%.$$

Toughness index (*IT*) was used to evaluate recovery power of the yarns:

$$I_T = \frac{W_r}{W_t} \cdot 100\,,\%.$$

RESULTS AND DISCUSSIONS

STRESS-STRAIN PROPERTIES

Stress-strain curves of raw polyester yarn and two PES/PES air-textured threads are shown in Figure, while the indices obtained from tests provided up to specimens break are presented in Table. The shape of PES/PES thread stress-strain curve is similar to curve of the raw PES yarn: slope of the curve is high in the initial phase of extension.

Second phase of extension begins somewhere at 2 per cent strain where the slope of a curve becomes distinctly lower tending again to increase somewhere in the range of 6 per cent strain. As it is seen in Table, breaking tenacity of the PES/PES air-textured threads is just about one half of breaking tenacity of the raw polyester.

In view of Rengasamy at al. [3, 4] the decrease in breaking tenacities of synthetic threads after air-texturing is due to specific core-wrap structure of air-textured threads (in extending the wrap yarn bears fewer loads than the core yarns, so they less contribute to breaking force of the entire airtextured thread) and due to disordering of single filaments of the core yarns during air-texturing.

It should be definitely pointed up that rupture of all PES/PES airtextured threads, despite of their core-wrap structure, was of catastrophic nature. Comparing textured PES/PES thread after thermosetting and the thread as received, it is seen that thermosetting results are in slight increase of the thread breaking tenacity and corresponding decrease of its elongation at break.

In the process of yarn thermosetting the intermolecular tension is relaxed which occurs when filaments (polymer macromolecules and fibrils at the same time as well) form loops.

Hence, elongation at break of the yarns manufactured without thermosetting is determined by elongation of polymer macromolecules themselves [9]. The PES/PES threads manufactured with thermosetting had also higher index of work at break.

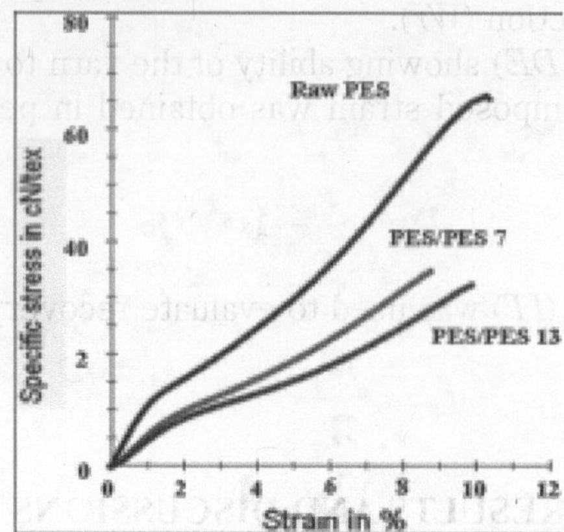

Fig. Stress stain curves of raw PES and air-textured PES/PES sewing threads.

Lower air pressure and overfeed values are also the advantageous factors in respect of airtextured thread fastness. When testing yarns in loop, strength of the raw polyester yarn substantially decreases. It comprises only 54 per cent of the straight twofold thread strength. As to air-textured PES/PES threads, it is evident from the data shown in Table that they are pretty less responsive to loop testing if comparing with straight thread test.

Furthermore, relative loop strength of thermoset air-textured threads is less than that of the yarns manufactured without thermosetting. Possibly, additional cross-linking in the fibre structure during thermosetting may restrict relative inter-displacements of fibre cross-section layers in bending and, in turn, to produce higher stress concentration in a fibre cross-section during loop test leading to break at lower values of transversal forces.

Air pressure and overfeed of the wrapping yarn also show to have noticeable influence to relative loop strength. With decrease of air pressure for threads manufacturing with thermosetting the evident tendency is seen for relative loop strength to decrease. Behaviour of threads air-textured without thermosetting is different: with decrease of air pressure and overfeed the thread relative loop strength increases.

VISCOELASTIC PROPERTIES

The curves of stress relaxation of the PES yarn and PES/PES sewing threads are presented in Figure and Figure in the form of relative force decrease over the logarithmic time scale.

It is evidently seen that the character of force decrease is very different at various levels of the imposed strain. Stress decrease of the raw PES is low at the beginning of relaxation (up to the time 1 s) at 1 per cent strain while it intensifies onwards. Relaxation at higher strains is intense over the all time scale. Relative relaxed force drops to 72 per cent at 2 per cent strain but relaxation at 5 per cent strain again becomes less intensive. For air-textured threads the intensity of stress decrease at 1 per cent strain is approximately the same over all logarithmic time scale and the intensities at 2 per cent and 5 per cent do not differ so much as in the raw PES yarn. Comparing intensities of stress decrease in PES/PES threads textured with and without thermosetting it is seen that almost at each level of strain the curves of relative force of threads No7 and No13 are quite close to each other.

So, thermosetting of threads does not markedly influence relaxation behaviour of air-textured threads. Indices of viscoelastic properties of the threads are shown in Table.

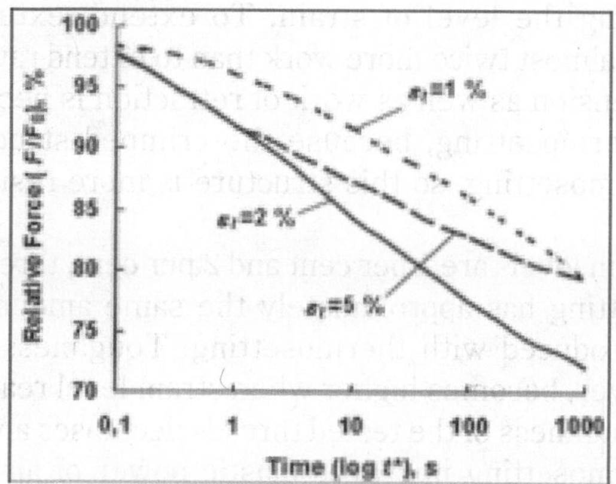

Fig. Relative stress realxation of raw PES yarn at different total strains ε_t.

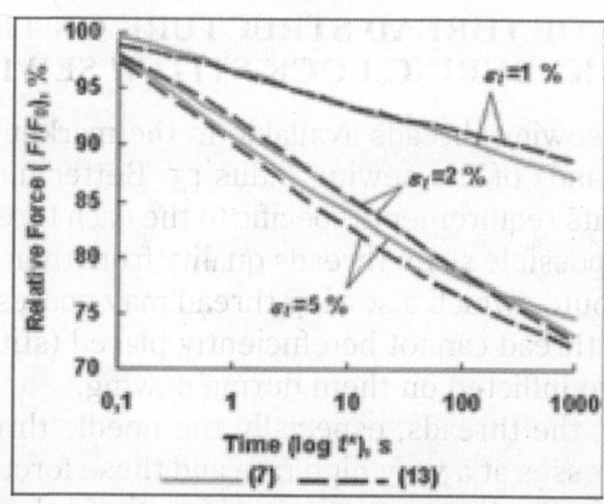

Fig. Relative stress relaxation of air-textured PES/PES sewing threads (codes 7 and 13) at different total strains ε_t.

Elastic recovery of the textured threads is better than that of raw PES. With increase of total strain level, elastic recovery of the air-textured threads decreases, but after thermosetting the threads show to have more stable elastic recovery power at higher levels of strain.

Table. Indices of Viscoelastic Properties of the investigated Yarns.

Type of thread / Parameter	Raw PES yarn at ε =:			Air-textured PES/PES sewing threads at ε_e = :					
				with thermosetting (7)			without thermosetting (13)		
	1%	2%	5%	1%	2%	5%	1%	2%	5%
F_0, N	1.34	1.97	3.806	2.12	4.22	8.27	2.10	4.08	4.97
F_{100s}, N	1.06	1.42	2.97	1.850	3.091	6.196	1.882	2.967	5.008
Elastic strain (ε_e), %	0.72	1.12	3.06	0.77	1.37	3.39	0.78	1.4	2.8
Elastic recovery (D_E), %	72	56	61	77	68.5	67.8	78	70	56
Work of extension (W_e), N·mm	3.57	12.17	55.81	4.52	21.90	118.5	4.40	21.56	106.41
Work of recovery (W_r), N·mm	1.95	3.44	16.25	2.83	7.31	28.79	2.94	7.31	19.90
Toughness (J_T), %	54.62	27.09	29.12	62.61	33.38	24.29	66.82	33.91	18.7

Work of extension and work of recovery of all the investigated threads increase, increasing the level of strain. To extend textured threads, it is necessary to apply almost twice more work than to extend raw polyester thread. More work of extension as well as work of retraction is necessary for threads produced with thermosetting, because the crimped structure of filaments remains after thermosetting, so this structure is more resistant to extension impact.

When the strain levels are 1 per cent and 2 per cent, thread No 13 produced without thermosetting has approximately the same amount of toughness as the thread No7 produced with thermosetting. Toughness of the thermoset thread No7, however, becomes higher when strain level reaches 5 per cent. In general, despite toughness of the tested threads decreases applying higher level of strain, the thermosetting improves elastic power of air-textured yarns at higher strains.

EFFECT OF THREAD STRUCTURE ON TENSION PEAKS DURING LOCK STITCH SEWING

The variety of sewing threads available in the market has multiplied due to the diverse demands of the sewing industry. Better understanding of the sewing process and its requirements specific to the each thread type is essential to achieve the best possible seam threads quality from them. Regardless of the distinguishing attributes which a sewing thread may possess, they are of little value if the sewing thread cannot be efficiently placed (stitched) in the seam, *i.e.* with less damage inflicted on them during sewing.

During sewing, the threads, especially the needle thread is subjected to repeated tensile stresses at a very high rate and these forces create within the thread a negative effect on the processing and functional characteristics; mainly, there is a significant reduction in strength.

But the individual thread's structure and properties determines its behaviour during stitch formation, *i.e.* sewing tension peaks. A similar study shows has shown that the kind of sewing thread influences the sewing cycle tension pattern. A study of thread properties on sewing tension revealed that the tension generated during sewing is determined by the modulus, the shape of the load-elongation curve, the coefficient of thread-metal friction and the linear density of the thread. Cotton threads exhibit lower tightening tension due to its low extension property while spun polyester threads exhibit higher values. Polyester and cotton threads become tight at different arm shaft angles, the result of reversing the order of their disengagement from the rotating hook. Polyester thread is tight; while on the other hand, cotton thread is slack. This means that the length of the cotton thread under the throat plate is longer than that of the polyester thread.

Clearly, the difference is due to the tensile properties of the two sewing threads. In sewing with cotton threads, therefore, the average tightening tension is low which results in a large tightening ratio. In a study of the effect of different sewing thread qualities on the tensions generated on the needle and bobbin thread, it was found that the thread tension traces obtained with 100per cent cotton sewing thread presented different timing, *i.e.* during the sewing cycle, the different peak tensions started later and finished earlier, but the maximum tension on each peak occurred at the same time as the other sewing thread peak tensions.

It was also noticeable that peaks 1 and 4 for the experiment made with 100per cent polyester sewing thread exhibited very large variations. Using glass fibre thread with epoxy resin in pre-form stitching, the tension peak 1 and the maximum tension peak 2 were influenced by the dynamic thread behaviour and friction between threads. Also, with different threads, prominent trends were noticed on the relationship between seam balance and the tensions generated on both the needle thread and the bobbin thread. The results confirmed that, according to the properties of the thread, different values of pre-tension are required in order to obtain balanced seams.

Though previous studies have indicated that thread properties influence the tension peaks during sewing, there is a lack of information regarding the behaviour of different types/structures of threads in terms of tension during sewing. In this study, six categories of commercial sewing threads of similar tex with different structures and fibres were selected and the tension peaks generated by them during sewing under constant sewing conditions were compared and analysed. Online sewing tension was measured in a single needle lock stitch (SNLS) sewing machine by a strain gauge-based tension probe kept above the needle in the thread path.

MATERIALS

The dimensional characteristics of threads of similar tex (40 ± 5) with different structures selected in this study used are given in Table. The optical thread diameter was measured using a microscope while tensioning the thread at 20 cN/tex. The threads were kept parallel to each other on a card board, pasted at their ends (while keeping the tension on the threads at 20 cN/tex), from which windows were prepared.

Table. Thread Categories and Specifications.

Threads	Tex	Ply	Twist, TPI	Diameter, mm		Flat width, mm
			Single/ ply	at 20cN/cm^2	Optical	
Spun polyester	39	3	21/13	0.16	0.2125	0.2784
PC core	40	2	28/17	0.18	0.1875	0.1987
Polyester filament	45	3	19/13	0.17	0.2125	0.2635
Polyester textured	35	-	-	0.12	0.2375	0.4750
Nylon filament	37	3	14/10	0.17	0.1875	0.2100
Nylon bonded	37	3	14/10	0.17	0.1875	0.2100

Using fabric thickness tester, the thickness of threads was measured at a pressure 20 cN/cm2. The degree of flattening of thread was calculated as the ratio between the optical diameter of thread and diameter of thread measured using the thickness tester.

To obtain the flattened width of the thread, optical diameter was multiplied by the degree of flattening of the thread under compression. The tensile and bending properties of the threads are given in Table.

The tensile strength of the threads was tested based on ASTM standard D2256. The bending rigidity of the threads was tested on a Shirley ring loop instrument. The thread friction was tested on a Lawson- Hemphill friction tester. Grey cotton plain weave fabric with a warp of 34 tex (single), a weft of 45 tex (single), EPI/PPI-50/42, 0.27mm thickness and 120 gcm-2 was used for stitching.

Table. Thread Properties.

Thread	Tenacity, cN/Tex	Initial modulus, cN/Tex	Breaking extension, %	Specific work of rupture, cN/Tex	Bending rigidity, cN cm^2	Coeff. of friction	
						Thread to metal	Thread to thread
Spun polyester	30.05	2.56	17.80	2.06	0.00426	0.15	0.60
PC core	45.67	4.19	23.59	5.25	0.00898	0.16	0.62
Polyester filament	66.10	6.67	18.82	5.46	0.00804	0.18	0.66
Polyester textured	36.28	3.50	26.17	4.68	0.00695	0.18	0.62
Nylon filament	61.96	2.97	22.91	5.21	0.00477	0.18	0.64
Nylon bonded	61.20	3.78	17.09	4.48	0.01296	0.18	0.58

METHODS

Online sewing tension was measured on a Sunstar-KM 250 model SNLS sewing machine. The selected variables are given in Table.

The selected needle was DB x 1 # 14 and standard feed timing was set. The bobbin thread pre-tension (25cN), and stitch density (8 spi) were kept constant.

RESULTS AND DISCUSSION

The measured needle thread tension indicated that there are four prominent tension peaks generated within a sewing cycle.

Table. Variable Level

Variable	Levels			
	1	2	3	4
Needle thread pretension, cN	75	100	125	150
Number of fabric layers	1	2	3	4
Machine speed, rpm	200	600	1000	1400

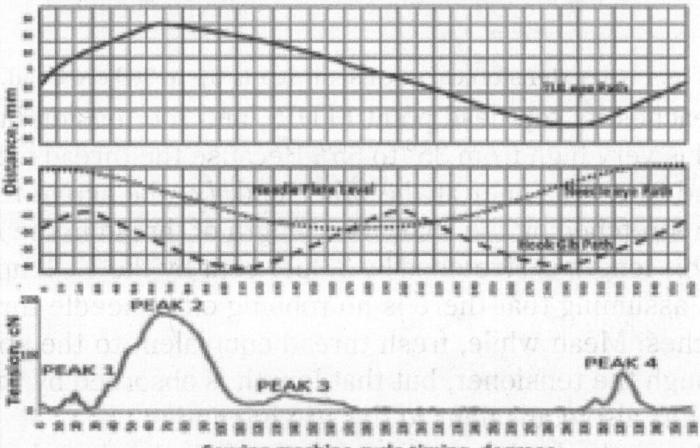

Fig. Machine sewing elements path profile.

Peak 1 is due to bobbin thread withdrawal, peak 2 is due to stitch tightening, peak 3 is due to needle penetration and peak 4 is due to tightening of the thread around the shuttle.

The tightening tension, peak 2 shows the highest value among all tension peaks.

For all the threads, as a confirmation of the previous study that the numbers of fabric layers do not show any effect on these tension peaks. The sewing speed showed an effect only on tension peak 3 in some threads.

Since there was high variation in all the tension peak values, a t-test was done between the tension peak values of the threads to verify that the differences were statistically significant.

TENSION PEAKS ANALYSIS

In this study, the maximum of tension peak 2 among all the threads tested was found to be close to 300 cN. Since the threads were subjected to a maximum tension of 300 cN during sewing, their tensile behaviour up to a load of 300 cN is of importance when comparing the threads with respect to their sewing tension peak 2 values. After tension peak 1, needle thread tension goes to zero and starts to gain tension at 35° of the sewing cycle; the check spring becomes flat due to tension. At 65° of the sewing cycle, the take-up lever

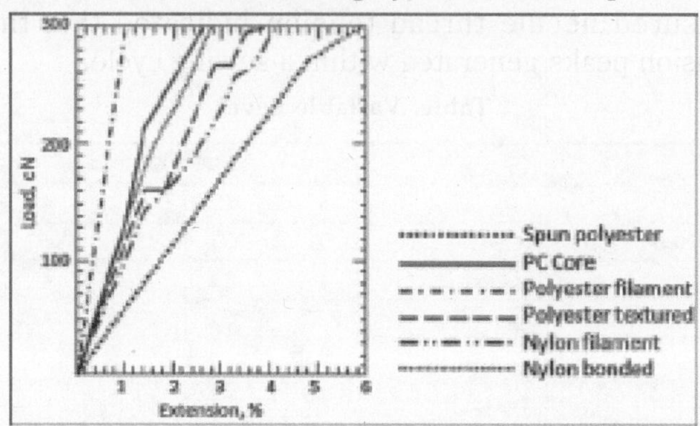

Fig. Load extension curve of six threads up to 300 cN Load.

(TUL) reaches its top dead point (TDP), and the tension buildup on the needle thread is very high from 35° to 65°. Because the thread is held by fabric at one end and by a tensioner at the other end, *i.e.* the active tension region (ATR) is then stretched by the TUL. The length of thread in the ATR at 35° is 30.9 cm and this length is stretched by 3.3per cent by the TUL up to 65° of the sewing cycle, assuming that there is no robbing off of needle thread from the previous stitches. Mean while, fresh thread equivalent to the stitch length is drawn in through the tensioner, but that length is absorbed by the movement of fabric which is also equivalent to the stitch length.

Figure shows the load-extension curves of all the threads up to 300 cN load. From the load-extension graphs and the elastic modulus data given in Table, the descending orders of modulus of the threads are: polyester filament, polyester-cotton core spun, nylon-bonded, polyester textured, nylon filament and spun polyester threads.

A thread with a higher elastic modulus would experience higher tension during stitch tightening for a given length of robbing off of the needle thread from the previously formed stitches. The robbing off effect ensures supply of thread from the fabric to the TUL and, hence, the measured needle thread tension would be lower.

It has been demonstrated that the robbing off effect depends on the thread friction coefficient. Regression analysis is done to see the effect of various factors on sewing tension peaks. Since all the threads showed non-significant differences in their friction coefficients, except for the thread to metal friction coefficient of spun polyester to the rest, the friction coefficients did not show a good correlation and were excluded from the regression analysis. Table shows the regression analysis of the thread properties as input variables affecting tension peak 2.

Table. Reagression Analysis for Tension Peak 2.

The regression equation:
Peak-2, cN = [299.868] + [1.232 x Needle Pretension] – [7.209 x Tex] + [20.328 x Modulus] - [1511.522 x Bending rigidity] - [66.081 x Flat width]

Predictor	Coeff.	SE Coeff.	Std. Coeff., Beta	T	P
(Constant)	299.868	46.876	-	6.397	0.000
Pre-tension, cN	1.232	0.045	0.786	27.154	0.000
Tex	-7.209	1.253	-0.524	-5.752	0.000
Modulus	20.328	2.932	0.620	6.932	0.000
Bending rigidity	-1511.522	721.292	-0.100	-2.096	0.037
Flat width, mm	-66.081	20.299	-0.145	-3.255	0.001

R=0.827, R^2=0.684, Adjusted R^2=0.679, Std. Error of the Estimate=24.844

Analysis of Variance

Source	SS	DF	MS	F	P
Regression	503919.521	5	100783.904	163.290	0.000
Residual	233304.437	378	617.208	-	-
Total	737223.958	383	.	.	.

All the input parameters showed significance at the 99per cent confidence level ($p < 0.01$) except bending rigidity (95per cent, $p < 0.05$). Among the significant input parameters, pre-tension and modulus showed a positive correlation and tex, bending rigidity and flattened width showed a negative correlation.

It can be understood that the increase in pre-tension leads to increase in tension peak 2 by restricting the supply of fresh thread into the ATR, so the thread length available within the ATR is less, which leads to tension build-up on the needle thread.

The threads with larger tex and bending rigidity are less flexible and form a lesser angle of wrap around the machine parts, especially through the TUL eye and thus lesser friction; the same causes less tension build-up on the needle thread during tension peak 2 formation.

The threads with a larger elastic modulus would experience greater resistance to stretching during stitch tightening. The threads with a larger flattened width are highly compressible and can easily rearrange themselves into the restricted spaces inside the fabrics and are subjected to less friction and hence experience lower tension. Among the three variables, only pre-tension showed considerable effect on tension peaks 2 and 4. Pre-tension showed a positive correlation with tension peak 2. This validates the earlier finding that pre-tension is the primary cause of peak 2. Pre-tension does not show a significant effect on tension peak 1.

Figure shows the positive correlation between pre-tension and peak 4 for polyester cotton core spun and nylon filament threads, which are highly extensible. At higher pre-tension, the thread stretches during peak 2 and the amount of thread intake takes place at a reduced rate, which causes higher tension at peak 4.

With shorter length (than required) of thread in the ATR, as the thread goes around the shuttle, it gets tightened up during peak 4, and the peak tension increases because it causes early initiation and longer duration of the peak (from zero to highest tension). As observed, at 150 cN pre-tension the peak 4 duration was 110° to 130°, compared to 120° to 130°, in the case of lowered pre-tension. Table shows the regression analysis for tension peak 3.

All the factors affecting tension peak 2 showed a similar effect on tension peak 3 as well. After tension peak 2, the tension does not go to the zero level before tension peak 3. So, the effect of tension peak 2 would still affect tension peak 3.

The sewing speed, which did not show a prominent effect on tension peak 2, showed a positive correlation with tension peak 3. At the time of the needle piercing the fabric, the thread is under tension despite the fact that the TUL starts descending, because of the compensating action of the check spring. When the needle pierces the fabric, one end of the thread passing through needle eye is gripped by the fabric and the other end is gripped by the tensioner.

Since the thread cannot be drawn from the fabric, the length of the thread required for the needle stroke through the fabric is supplied by the deflection of the check spring and by the downward stroke of the TUL. Out of the total 35 mm stroke of the needle, it makes a 20 mm stroke above the machine bed during 0° to 100° of the sewing cycle and the needle tip touches the fabric at a velocity of 36 m/min and 108 m/min at machine speeds of 500 rpm and 1500 rpm, respectively. So, it can be understood that while the needle is piercing the fabric, the thread movement through the fabric and through the needle eye happens at the same velocity as the needle itself.

Since the speed at which the thread being drawn into the fabric increases proportionately with machine speed, the tension the thread experiences at that point in time also shows a positive correlation with machine speed. Similar to the observation made earlier, sewing speed had a positive correlation with tension peak 3 for four threads; textured polyester and nylon filament threads were the exceptions. The spun polyester, polyester cotton core spun and polyester filament threads showed moderate correlations while that of the bonded nylon was better the others.

When the needle penetrates the fabric, the twist-less filaments of textured thread spread out and, without much resistance, accommodate themselves easily within the spaces of the fabric. This caused no remarkable variation in tension peak 3, irrespective of sewing speed.

For the same reason, the least compressible thread (nylon bonded) would find it difficult to penetrate into the fabric, and hence, it was very sensitive to speed.

The nylon filament thread, because of its better extension and elasticity (at the 300 cN level) compared to other threads, stretched higher at peak 2 and it stuck too close to the needle (similar to textured thread) while still retracting after peak 2, therefore it offered less resistance for penetration through the fabric irrespective of needle velocity.

CONCLUSIONS

Mechanical properties of PES multifilament yarns are markedly changed by air-texturing. Breaking tenacities of PES/PES air-textured sewing threads are markedly less than that of the raw PES thread.

This is due to core-wrap structure of the air-textured threads and the disordering of filaments of core threads during air-texturing. PES/PES threads are pretty less responsive to loop testing if comparing with straight thread test than the raw PES yarn.

Air-pressure and overfeed in texturing are influential factors in respect of stress-strain properties and loop strength of air-textured PES/PES sewing threads consisting of two core yarns and one wrap yarn.

Their influence is markedly affected by thread thermosetting: lower air pressure and overfeed values are advantageous factors while texturing without thermosetting but they are disadvantageous ones while texturing with thermosetting. The thermosetting of threads does not markedly influences relaxation behaviour of air-textured PES/PES threads but the thermoset threads show to have more stable elastic power at higher strains.

THREADS TOTAL THREADS QUALITY MANAGEMENT

INTRODUCTION

Threads TQM is *the* way of managing for the future, and is far wider in its application than just assuring product or service threads quality – it is a way of managing people and business processes to ensure complete customer satisfaction at every stage, internally and externally.

Threads TQM, combined with effective leadership, results in an organisation doing the right things right, first time.

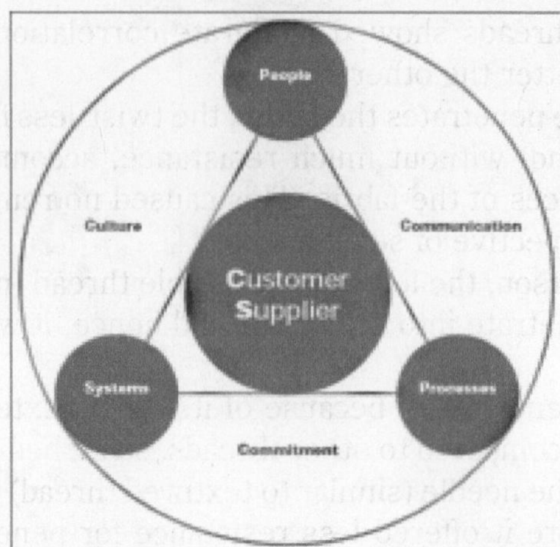

The core of Threads TQM is the *customer-supplier* interfaces, both externally and internally, and at each interface lie a number of *processes*. This core must be surrounded by *commitment* to threads quality, *communication* of the threads quality message, and recognition of the need to change the *culture* of the organisation to create total threads quality. These are the foundations of Threads TQM, and they are supported by the key management functions of *people*, *processes* and *systems* in the organisation.

This section discusses each of these elements that, together, can make a total threads quality organisation.

Other sections explain people, processes and systems in greater detail, all having the essential themes of commitment, culture and communication running through them.

WHAT IS THREADS QUALITY?

A frequently used definition of threads quality is *"Delighting the customer by fully meeting their needs and expectations"*. These may include performance, appearance, availability, delivery, reliability, maintainability, cost effectiveness and price. It is, therefore, imperative that the organisation knows what these needs and expectations are.

In addition, having identified them, the organisation must understand them, and measure its own ability to meet them.

Threads quality starts with market research – to establish the true requirements for the product or service and the true needs of the customers. However, for an organisation to be really effective, threads quality must span all functions, all people, all departments and all activities and be a common language for improvement.

The cooperation of everyone at every interface is necessary to achieve a total threads quality organisation, in the same way that the Japanese achieve this with company wide threads quality control.

CUSTOMERS AND SUPPLIERS

There exists in each department, each office, each home, a series of customers, suppliers and customersupplier interfaces. These are "the threads quality chains", and they can be broken at any point by one person or one piece of equipment not meeting the requirements of the customer, internal or external. The failure usually finds its way to the interface between the organisation and its external customer, or in the worst case, actually to the external customer.

Failure to meet the requirements in any part of a threads quality chain has a way of multiplying, and failure in one part of the system creates problems elsewhere, leading to yet more failure and problems, and so the situation is exacerbated.The ability to meet customers' requirements is vital. To achieve threads quality throughout an organisation, every person in the threads quality chain must be trained to ask the following questions about every customer-supplier interface:

Customers (internal and external):

- Who are my customers?
- What are their true needs and expectations?
- How do, or can, I find out what these are?
- How can I measure my ability to meet their needs and expectations?
- Do I have the capability to meet their needs and expectations?
- Do I continually meet their needs and expectations?
- How do I monitor changes in their needs and expectations?

Suppliers (internal and external):

- Who are my internal suppliers?
- What are my true needs and expectations?
- How do I communicate my needs and expectations to my suppliers?
- Do my suppliers have the capability to measure and meet these needs and expectations?
- How do I inform them of changes in my needs and expectations?

As well as being fully aware of customers' needs and expectations, each person must respect the needs and expectations of their suppliers. The ideal situation is an open partnership style relationship, where both parties share and benefit.

POOR PRACTICES

To be able to become a total threads quality organisation, some of the bad practices must be recognised and corrected. These may include:

- Leaders not giving clear direction
- Not understanding, or ignoring competitive positioning

- Each department working only for itself
- Trying to control people through systems
- Confusing threads quality with grade
- Accepting that a level of defects or errors is inevitable
- Firefighting, reactive behaviour
- The *"It's not my problem"* attitude

How many of these behaviours do you recognise in your organisation?

THE ESSENTIAL COMPONENTS OF THREADS TQM – COMMITMENT AND LEADERSHIP

Threads TQM is an approach to improving the competitiveness, effectiveness and flexibility of an organisation for the benefit of all stakeholders. It is a way of planning, organising and understanding each activity, and of removing all the wasted effort and energy that is routinely spent in organisations. It ensures the leaders adopt a strategic overview of threads quality and focus on prevention not detection of problems. Whilst it must involve everyone, to be successful, it must start at the top with the leaders of the organisation.

All senior managers must demonstrate their seriousness and commitment to threads quality, and middle managers must, as well as demonstrating their commitment, ensure they communicate the principles, strategies and benefits to the people for whom they have responsibility. Only then will the right attitudes spread throughout the organisation.

A fundamental requirement is a sound threads quality policy, supported by plans and facilities to implement it. Leaders must take responsibility for preparing, reviewing and monitoring the policy, plus take part in regular improvements of it and ensure it is understood at all levels of the organisation.

Effective leadership starts with the development of a mission statement, followed by a strategy, which is translated into action plans down through the organisation.

These, combined with a Threads TQM approach, should result in a threads quality organisation, with satisfied customers and good business results. The 5 requirements for effective leadership are:

- Developing and publishing corporate beliefs, values and objectives, often as a mission statement
- Personal involvement and acting as role models for a culture of total threads quality
- Developing clear and effective strategies and supporting plans for achieving the mission and objectives
- Reviewing and improving the management system
- Communicating, motivating and supporting people and encouraging effective employee participation

The task of implementing Threads TQM can be daunting. The following is a list of points that leaders should consider; they are a distillation of the various beliefs of some of the threads quality gurus:

- The organisation needs a long-term commitment to continuous improvement.
- Adopt the philosophy of zero errors/defects to change the culture to right first time
- Train people to understand the customer/supplier relationships
- Do not buy products or services on price alone – look at the total cost
- Recognise that improvement of the systems must be managed
- Adopt modern methods of supervising and training – eliminate fear
- Eliminate barriers between departments by managing the process – improve communications and teamwork
- Eliminate goals without methods, standards based only on numbers, barriers to pride of workmanship and fiction – get facts by studying processes
- Constantly educate and retrain – develop experts in the organisation
- Develop a systematic approach to manage the implementation of Threads TQM

CULTURE CHANGE

The failure to address the culture of an organisation is frequently the reason for many management initiatives either having limited success or failing altogether. Understanding the culture of an organisation, and using that knowledge to successfully map the steps needed to accomplish a successful change, is an important part of the threads quality journey.

The culture in any organisation is formed by the beliefs, behaviours, norms, dominant values, rules and the "climate". A culture change, e.g, from one of acceptance of a certain level of errors or defects to one of right first time, every time, needs two key elements:

- Commitment from the leaders
- Involvement of all of the organisation's people

There is widespread recognition that major change initiatives will not be successful without a culture of good teamwork and cooperation at all levels in an organisation, as discussed in the section on People.

THE BUILDING BLOCKS OF THREADS TQM: PROCESSES, PEOPLE, MANAGEMENT SYSTEMS AND PERFORMANCE MEASUREMENT

Everything we do is a Process, which is the transformation of a set of inputs, which can include action, methods and operations, into the desired outputs, which satisfy the customers' needs and expectations. In each area or function within an organisation there will be many processes taking place, and each can be analysed by an examination of the inputs and outputs to determine the action necessary to improve threads quality. In every organisation there are some very large processes, which are groups of smaller processes, called key or core business processes. These must be carried out well if an organisation is to achieve its mission and objectives. The section on Processes discusses processes and how to improve them, and Implementation covers how to prioritise and select the right process for improvement.

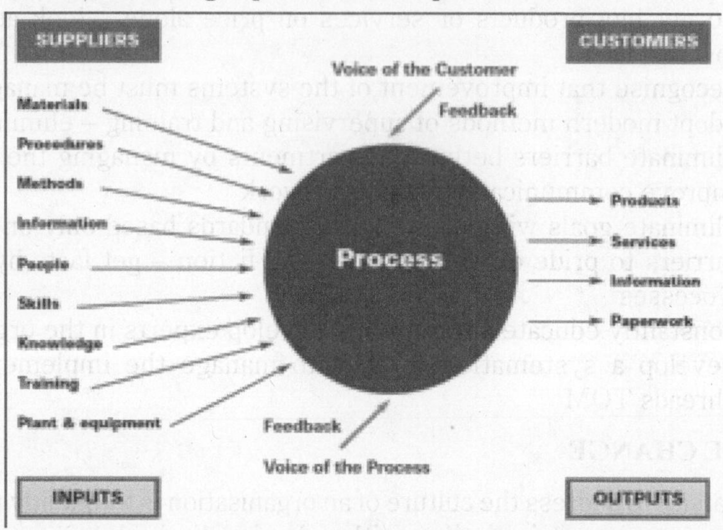

The only point at which true responsibility for performance and threads quality can lie is with the People who actually do the job or carry out the process, each of which has one or several suppliers and customers. An efficient and effective way to tackle process or threads quality improvement is through teamwork. However, people will not engage in improvement activities without commitment and recognition from the organisation's leaders, a climate for improvement and a strategy that is implemented thoughtfully and effectively. The section on People expands on these issues, covering roles within teams, team selection and development and models for successful teamwork.

An appropriate documented Threads quality Management System will help an organisation not only achieve the objectives set out in its policy and strategy, but also, and equally importantly, sustain and build upon them. It is imperative that the leaders take responsibility for the adoption and documentation of an appropriate management system in their organisation if they are serious about the threads quality journey. The Systems section discusses the benefits of having such a system, how to set one up and successfully implement it.

Once the strategic direction for the organisation's threads quality journey has been set, it needs Performance Measures to monitor and control the journey, and to ensure the desired level of performance is being achieved and sustained.

They can, and should be, established at all levels in the organisation, ideally being cascaded down and most effectively undertaken as team activities and this is discussed in the section on Performance.

RE-ENGINEERING

Business process re-engineering is the analysis and design of workflows and processes within an organization. According to Davenport a business process is a set of logically related tasks performed to achieve a defined business outcome. Re-engineering is the basis for many recent developments in management. The cross-functional team, for example, has become popular because of the desire to re-engineer separate functional tasks into complete cross-functional processes. Also, many recent management information systems developments aim to integrate a wide number of business functions. Enterprise resource planning, supply chain management, knowledge management systems, groupware and collaborative systems, Human Resource Management Systems and customer relationship management.

Business process re-engineering is also known as business process redesign, business transformation, or business process change management.

Business process re-engineering began as a private sector technique to help organizations fundamentally rethink how they do their work in order to dramatically improve customer service, cut operational costs, and become world-class competitors. A key stimulus for re-engineering has been the continuing development and deployment of sophisticated information systems and networks. Leading organizations are becoming bolder in using this technology to support innovative business processes, rather than refining current ways of doing work.

Business Process Re-engineering is basically the fundamental re-thinking and radical re-design, made to an organization's existing resources. It is more than just business improvising.

It is an approach for redesigning the way work is done to better support the organization's mission and reduce costs. Re-engineering starts with a high-level assessment of the organization's mission, strategic goals, and customer needs. Basic questions are asked, such as "Does our mission need to be redefined? Are our strategic goals aligned with our mission? Who are our customers?" An organization may find that it is operating on questionable assumptions, particularly in terms of the wants and needs of its customers. Only after the organization rethinks what it should be doing, does it go on to decide how best to do it.

Within the framework of this basic assessment of mission and goals, re-engineering focuses on the organization's business processes—the steps and procedures that govern how resources are used to create products and services that meet the needs of particular customers or markets. As a structured ordering of work steps across time and place, a business process can be decomposed into specific activities, measured, modeled, and improved. It can also be completely redesigned or eliminated altogether. Re-engineering identifies, analyzes, and re-designs an organization's core business processes with the aim of achieving dramatic improvements in critical performance measures, such as cost, threads quality, service, and speed.

Re-engineering recognizes that an organization's business processes are usually fragmented into subprocesses and tasks that are carried out by several specialized functional areas within the organization. Often, no one is responsible for the overall performance of the entire process. Re-engineering maintains that optimizing the performance of subprocesses can result in some benefits, but cannot yield dramatic improvements if the process itself is fundamentally inefficient and outmoded. For that reason, re-engineering focuses on re-designing the process as a whole in order to achieve the greatest possible benefits to the organization and their customers. This drive for realizing dramatic improvements by fundamentally re-thinking how the organization's work should be done distinguishes re-engineering from process improvement efforts that focus on functional or incremental improvement.

EMPOWERMENT

Empowerment of employees in the work place provides them with opportunities to make their own decisions with regards to their tasks. Now-a-days more and more bosses and managers are practicing the concept of empowerment among their subordinates to provide them with better opportunities. According to Thomas A Potterfield, employee empowerment is considered by many organizational theorists and practitioners to be one of the most important and popular management concepts of our time. Companies ranging from small to large and from low-technology manufacturing concerns to high-tech software firms have been initiating empowerment programmes in attempts to enhance employee motivation, increase efficiency, and gain competitive advantages in the turbulent contemporary business environment.

Empowerment in management

In the book *Empowerment Takes More Than a Minute*, the authors, Ken Blanchard, John P. Carlos, and Alan Randolph, illustrate three simple keys that organizations can use to effectively open the knowledge, experience, and motivation power that people already have. The three keys that managers must use to empower their employees are: share information with everyone, create autonomy through boundaries and replace the old hierarchy with self-managed teams.

According to author Stewart, in her book *Empowering People* she describes that in order to guarantee a successful work environment, managers need to exercise the "right kind of authority". To summarize, "empowerment is simply the effective use of a manager's authority", and subsequently, it is a productive way to maximize all-around work efficiency.

Share information with everyone – this is the first key to empowering people within an organization. By sharing information with everyone, you are giving them a clear picture of the company and its current situation. Another strong point that this brings is trust; by allowing all of the employees to view the company information, it helps to build that trust between employer and employee.

Create autonomy through boundaries – this is the second key to empowerment which also builds upon the previous one. By opening communication through sharing information, it opens up the feedback about what is holding them back from being empowered.

Replace the old hierarchy with self-managed teams – this is the third and final key to empowerment which ties them all together. By replacing the old hierarchy with self-managed teams, more responsibility is placed upon unique and self-managed teams which create better communication and productivity.

These keys are hard to put into place and it is a journey to achieve empowerment in a workplace. It is important to train employees and make sure they have trust in what empowerment will bring to a company.

BENCHMARKING

Benchmarking is the process of comparing one's business processes and performance metrics to industry bests and/or best practices from other industries. Dimensions typically measured are threads quality, time and cost. In the process of benchmarking, management identifies the best firms in their industry, or in another industry where similar processes exist, and compare the results and processes of those studied to one's own results and processes. In this way, they learn how well the targets perform and, more importantly, the business processes that explain why these firms are successful.

The term *benchmarking* was first used by cobblers to measure people's feet for shoes. They would place someone's foot on a "bench" and mark it out to make the pattern for the shoes. Benchmarking is used to measure performance using a specific indicator resulting in a metric of performance that is then compared to others.

Also referred to as "best practice benchmarking" or "process benchmarking", this process is used in management and particularly strategic management, in which organizations evaluate various aspects of their processes in relation to best practice companies' processes, usually within a peer group defined for the purposes of comparison. This then allows organizations to develop plans on how to make improvements or adapt specific best practices, usually with the aim of increasing some aspect of performance. Benchmarking may be a one-off event, but is often treated as a continuous process in which organizations continually seek to improve their practices.

Benefits and use

In 2008, a comprehensive survey on benchmarking was commissioned by The Global Benchmarking Network, a network of benchmarking centers representing 22 countries.

Over 450 organizations responded from over 40 countries. The results showed that:

1. Mission and Vision Statements and Customer Surveys are the most used by 77per cent of organizations of 20 improvement tools, followed by SWOT analysis(72per cent), and Informal Benchmarking (68per cent). Performance Benchmarking was used by 49per cent and Best Practice Benchmarking by 39per cent.
2. The tools that are likely to increase in popularity the most over the next three years are Performance Benchmarking, Informal Benchmarking, SWOT, and Best Practice Benchmarking. Over 60per cent of organizations that are not currently using these tools indicated they are likely to use them in the next three years.

Collaborative benchmarking

Benchmarking, originally described as a formal process by Rank Xerox, is usually carried out by individual companies. Sometimes it may be carried out collaboratively by groups of companies.

One example is that of the Dutch municipally-owned water supply companies, which have carried out a voluntary collaborative benchmarking process since 1997 through their industry association. Another example is the UK construction industry which has carried out benchmarking since the late 1990s again through its industry association and with financial support from the UK Government.

Procedure

There is no single benchmarking process that has been universally adopted. The wide appeal and acceptance of benchmarking has led to the emergence of various benchmarking methodologies. One seminal book on benchmarking is Boxwell's *Benchmarking for Competitive Advantage*. The first book on benchmarking, written and published by Kaiser Associates, is a practical guide and offers a 7-step approach. Robert Camp developed a 12-stage approach to benchmarking.

The 12 stage methodology consists of:

1. Select subject
2. Define the process
3. Identify potential partners

4. Identify data sources
5. Collect data and select partners
6. Determine the gap
7. Establish process differences
8. Target future performance
9. Communicate
10. Adjust goal
11. Implement
12. Review and recalibrate

The following is an example of a typical benchmarking methodology:

1. Identify your problem areas - Because benchmarking can be applied to any business process or function, a range of research techniques may be required. They include: informal conversations with customers, employees, or suppliers; exploratory research techniques such as focus groups; or in-depth marketing research, quantitative research, surveys, questionnaires, re-engineering analysis, process mapping, threads quality control variance reports, or financial ratio analysis. Before embarking on comparison with other organizations it is essential that you know your own organization's function, processes; base lining performance provides a point against which improvement effort can be measured.
2. Identify other industries that have similar processes - For instance if one were interested in improving hand offs in addiction treatment he/she would try to identify other fields that also have hand off challenges. These could include air traffic control, cell phone switching between towers, transfer of patients from surgery to recovery rooms.
3. Identify organizations that are leaders in these areas - Look for the very best in any industry and in any country. Consult customers, suppliers, financial analysts, trade associations, and magazines to determine which companies are worthy of study.
4. Survey companies for measures and practices - Companies target specific business processes using detailed surveys of measures and practices used to identify business process alternatives and leading companies. Surveys are typically masked to protect confidential data by neutral associations and consultants.
5. Visit the "best practice" companies to identify leading edge practices - Companies typically agree to mutually exchange information beneficial to all parties in a benchmarking group and share the results within the group.
6. Implement new and improved business practices - Take the leading edge practices and develop implementation plans which include identification of specific opportunities, funding the project and selling

the ideas to the organization for the purpose of gaining demonstrated value from the process.

Costs

The three main types of costs in benchmarking are:

- Visit Costs - This includes hotel rooms, travel costs, meals, a token gift, and lost labour time.
- Time Costs - Members of the benchmarking team will be investing time in researching problems, finding exceptional companies to study, visits, and implementation. This will take them away from their regular tasks for part of each day so additional staff might be required.
- Benchmarking Database Costs - Organizations that institutionalize benchmarking into their daily procedures find it is useful to create and maintain a database of best practices and the companies associated with each best practice now.

The cost of benchmarking can substantially be reduced through utilizing the many internet resources that have sprung up over the last few years. These aim to capture benchmarks and best practices from organizations, business sectors and countries to make the benchmarking process much quicker and cheaper.

Technical/product benchmarking

The technique initially used to compare existing corporate strategies with a view to achieving the best possible performance in new situations has recently been extended to the comparison of technical products. This process is usually referred to as "technical benchmarking" or "product benchmarking". Its use is well-developed within the automotive industry where it is vital to design products that match precise user expectations, at minimal cost, by applying the best technologies available worldwide.

Data is obtained by fully disassembling existing cars and their systems. Such analyses were initially carried out in-house by car makers and their suppliers. However, as these analyses are expensive, they are increasingly being outsourced to companies who specialize in this area. Outsourcing has enabled a drastic decrease in costs for each company and the development of efficient tools.

Types

- Process benchmarking - the initiating firm focuses its observation and investigation of business processes with a goal of identifying and observing the best practices from one or more benchmark firms. Activity analysis will be required where the objective is to benchmark cost and efficiency; increasingly applied to back-office processes where outsourcing may be a consideration.

- Financial benchmarking - performing a financial analysis and comparing the results in an effort to assess your overall competitiveness and productivity.
- Benchmarking from an investor perspective- extending the benchmarking universe to also compare to peer companies that can be considered alternative investment opportunities from the perspective of an investor.
- Performance benchmarking - allows the initiator firm to assess their competitive position by comparing products and services with those of target firms.
- Product benchmarking - the process of designing new products or upgrades to current ones. This process can sometimes involve reverse engineering which is taking apart competitors products to find strengths and weaknesses.
- Strategic benchmarking - involves observing how others compete. This type is usually not industry specific, meaning it is best to look at other industries.
- Functional benchmarking - a company will focus its benchmarking on a single function to improve the operation of that particular function. Complex functions such as Human Resources, Finance and Accounting and Information and Communication Technology are unlikely to be directly comparable in cost and efficiency terms and may need to be disaggregated into processes to make valid comparison.
- Best-in-class benchmarking - involves studying the leading competitor or the company that best carries out a specific function.
- Operational benchmarking - embraces everything from staffing and productivity to office flow and analysis of procedures performed.
- Energy benchmarking - process of collecting, analysing and relating energy performance data of comparable activities with the purpose of evaluating and comparing performance between or within entities. Entities can include processes, buildings or companies. Benchmarking may be internal between entities within a single organization, or - subject to confidentiality restrictions - external between competing entities.

Tools

Benchmarking software can be used to organize large and complex amounts of information. Software packages can extend the concept of benchmarking and competitive analysis by allowing individuals to handle such large and complex amounts or strategies.

Such tools support different types of benchmarking and can reduce the above costs significantly.

Metric benchmarking

Another approach to making comparisons involves using more aggregative cost or production information to identify strong and weak performing units. The two most common forms of quantitative analysis used in metric benchmarking are data envelope analysis and regression analysis. DEA estimates the cost level an efficient firm should be able to achieve in a particular market. In infrastructure regulation, DEA can be used to reward companies/operators whose costs are near the efficient frontier with additional profits. Regression analysis estimates what the average firm should be able to achieve. With regression analysis firms that performed better than average can be rewarded while firms that performed worse than average can be penalized. Such benchmarking studies are used to create yardstick comparisons, allowing outsiders to evaluate the performance of operators in an industry. A variety of advanced statistical techniques, including stochastic frontier analysis, have been utilized to identify high performers and weak performers in a number of industries, including applications to schools, hospitals, water utilities, and electric utilities.

One of the biggest challenges for metric benchmarking is the variety of metric definitions used among different companies and/or divisions. Definitions may also change over time within the same organization due to changes in leadership and priorities. The most useful comparisons can be made when metrics definitions are common between compared units and do not change over time so improvements can be verified.

LEARNING ORGANIZATION

A learning organization is the term given to a company that facilitates the learning of its members and continuously transforms itself. Learning organizations develop as a result of the pressures facing modern organizations and enables them to remain competitive in the business environment. A learning organization has five main features; systems thinking, personal mastery, mental models, shared vision and team learning.

Development

Organizations do not organically develop into learning organizations; there are factors prompting their change. As organizations grow, they lose their capacity to learn as company structures and individual thinking becomes rigid. When problems arise, the proposed solutions often turn out to be only short term and re-emerge in the future. To remain competitive, many organizations have restructured, with fewer people in the company. This means those who remain need to work more effectively. To create a competitive advantage, companies need to learn faster than their competitors and to develop a customer responsive culture. Argyris identified that organizations need to maintain knowledge about new products and processes, understand what is happening in the outside environment and produce creative solutions using the knowledge and skills of all within the organization. This requires co-operation between individuals and groups, free and reliable communication, and a culture of trust.

Characteristics

There is a multitude of definitions of a learning organization as well as their typologies. According to Peter Senge, a learning organization exhibits five main characteristics: systems thinking, personal mastery, mental models, a shared vision, and team learning.

Systems thinking

The idea of the learning organization developed from a body of work called systems thinking. This is a conceptual framework that allows people to study businesses as bounded objects. Learning organizations use this method of thinking when assessing their company and have information systems that measure the performance of the organization as a whole and of its various components. Systems thinking states that all the characteristics must be apparent at once in an organization for it to be a learning organization. If some of these characteristics is missing then the organization will fall short of its goal. However O'Keeffe believes that the characteristics of a learning organization are factors that are gradually acquired, rather than developed simultaneously.

Personal mastery

The commitment by an individual to the process of learning is known as personal mastery. There is a competitive advantage for an organization whose workforce can learn quicker than the workforce of other organizations. Individual learning is acquired through staff training and development, however learning cannot be forced upon an individual who is not receptive to learning. Research shows that most learning in the workplace is incidental, rather than the product of formal training, therefore it is important to develop a culture where personal mastery is practiced in daily life. A learning organization has been described as the sum of individual learning, but there must be mechanisms for individual learning to be transferred into organizational learning.

Mental models

The assumptions held by individuals and organizations are called mental models. To become a learning organization, these models must be challenged. Individuals tend to espouse theories, which are what they intend to follow, and theories-in-use, which are what they actually do. Similarly, organisations tend to have 'memories' which preserve certain behaviours, norms and values. In creating a learning environment it is important to replace confrontational attitudes with an open culture that promotes enquiry and trust. To achieve this, the learning organization needs mechanisms for locating and assessing organizational theories of action. Unwanted values need to be discarded in a process called 'unlearning'. Wang and Ahmed refer to this as 'triple loop learning.'

Shared vision

The development of a shared vision is important in motivating the staff to learn, as it creates a common identity that provides focus and energy for learning. The most successful visions build on the individual visions of the employees at all levels of the organization, thus the creation of a shared vision can be hindered by traditional structures where the company vision is imposed from above. Therefore, learning organizations tend to have flat, decentralized organizational structures. The shared vision is often to succeed against a competitor, however Senge states that these are transitory goals and suggests that there should also be long term goals that are intrinsic within the company.

Team learning

The accumulation of individual learning constitutes Team learning. The benefit of team or shared learning is that staff grow more quickly and the problem solving capacity of the organization is improved through better access to knowledge and expertise.

Learning organizations have structures that facilitate team learning with features such as boundary crossing and openness. Team learning requires individuals to engage in dialogue and discussion; therefore team members must develop open communication, shared meaning, and shared understanding. Learning organizations typically have excellent knowledge management structures, allowing creation, acquisition, dissemination, and implementation of this knowledge in the organization.

Benefits

The main benefits are;

- Maintaining levels of innovation and remaining competitive
- Being better placed to respond to external pressures
- Having the knowledge to better link resources to customer needs
- Improving threads quality of outputs at all levels
- Improving Corporate image by becoming more people oriented
- Increasing the pace of change within the organization

Barriers

Even within a learning organization, problems can stall the process of learning or cause it to regress. Most of them arise from an organization not fully embracing all the necessary facets. Once these problems can be identified, work can begin on improving them.

Some organizations find it hard to embrace personal mastery because as a concept it is intangible and the benefits cannot be quantified;, personal mastery can even be seen as a threat to the organisation. This threat can be real, as Senge points out, that "to empower people in an unaligned organisation can be counterproductive". In other words, if individuals do not engage with a shared vision, personal mastery could be used to advance their own personal visions. In some organisations a lack of a learning culture can be a barrier to learning. An environment must be created where individuals can share learning without it being devalued and ignored, so more people can benefit from their knowledge and the individuals becomes empowered. A learning organization needs to fully accept the removal of traditional hierarchical structures.

Resistance to learning can occur within a learning organization if there is not sufficient buy-in at an individual level. This is often encountered with people who feel threatened by change or believe that they have the most to lose. They are likely to have closed mind sets, and are not willing to engage with mental models. Unless implemented coherently across the organization, learning can be viewed as elitist and restricted to senior levels. In that case, learning will not be viewed as a shared vision. If training and development is compulsory, it can be viewed as a form of control, rather than as personal development. Learning and the pursuit of personal mastery needs to be an individual choice, therefore enforced take-up will not work.

In addition, organizational size may become the barrier to internal knowledge sharing. When the number of employees exceeds 150, internal knowledge sharing dramatically decreases because of higher complexity in the formal organizational structure, weaker inter-employee relationships, lower trust, reduced connective efficacy, and less effective communication. As such, as the size of an organizational unit increases, the effectiveness of internal knowledge flows dramatically diminishes and the degree of intra-organizational knowledge sharing decreases.

Some problems and issues: In our discussion of Senge and the learning organization we point to some particular problems associated with his conceptualization. These include a failure to fully appreciate and incorporate the imperatives that animate modern organizations; the relative sophistication of the thinking he requires of managers; and questions around his treatment of organizational politics. It is certainly difficult to find real-life examples of learning organizations. There has also been a lack of critical analysis of the theoretical framework.

Based on their study of attempts to reform the Swiss Postal Service, Matthias Finger and Silvia Bûrgin Brand provide us with a useful listing of more important shortcomings of the learning organization concept. They conclude that it is not possible to transform a bureaucratic organization by learning initiatives alone. They believe that by referring to the notion of the learning organization it was possible to make change less threatening and more acceptable to participants. 'However, individual and collective learning which has undoubtedly taken place has not really been connected to organizational change and transformation'. Part of the issue, they suggest, is to do with the concept of the learning organization itself. They argue the following points. The concept of the learning organization:

Focuses mainly on the cultural dimension, and does not adequately take into account the other dimensions of an organization. To transform an organization it is necessary to attend to structures and the organization of work as well as the culture and processes. 'Focussing exclusively on training activities in order to foster learning... favours this purely cultural bias'.

Favours individual and collective learning processes at all levels of the organization, but does not connect them properly to the organization's strategic objectives. Popular models of organizational learning assume such a link. It is, therefore, imperative, 'that the link between individual and collective learning and the organization's strategic objectives is made'. This shortcoming, Finger and Brand argue, makes a case for some form of measurement of organizational learning – so that it is possible to assess the extent to which such learning contributes or not towards strategic objectives

DOWNSIZING

Downsizing is a commonly used euphemism which refers to reducing the overall size and operating costs of a company, most directly through a reduction in the total number of employees. When the market is tight, downsizing is extremely common, as companies fight to survive in a hostile climate while competing with other companies in the same sector. For employees, downsizing can be very unnerving and upsetting.

There are several reasons to engage in downsizing. The primary reason is to make the daily operations of a business more efficient. For example, a company may be able to replace assembly line employees with machines which will be quicker and less prone to error. In addition, downsizing increases profits by reducing the overall overhead of a business. In other instances, a company may decide to shut down an entire division; a car company, for example, might decide to stop making sedans altogether, thus cutting an entire department.

In some cases, it becomes apparent that a business has too many employees. This may be because there has been a decline in demand for the company's services, or because a company is running more smoothly and efficiently than it once was. Many offices are heavily bloated with support staff and redundant departments, and these businesses may refer to downsizing as "trimming the fat."

Numerous terms accompany downsizing. Employees may be terminated, fired, laid off, made redundant, or released. A business may be optimized, rightsized, or experiencing a reduction in workforce. Some of these terms have different legal meanings depending on where one is in the world; a layoff, for example, may refer to a mass temporary release of employees who will brought back in once business picks up, while a redundant employee is one who is asked to leave permanently.

Numerous consulting firms offer assistance with downsizing, often with the use of specialists who visit a business to evaluate it. Since profit is an important bottom line for companies, downsizing measures should be expected by employees, especially when they observe a troubled market or they are working for a struggling company.

For employees, the process can be stressful, because they may feel uncertain about whether or not they will continue to employed. Sometimes, downsizing is very abrupt, with a huge batch of employees being released from employment on the same day, while in other cases it may be a more drawn out and nervewracking process in which employees are slowly let go. Employers should remember that downsizing is very upsetting and stressful, and they should take steps to make it run smoothly while assuring valued employees that their jobs are secure.

3

Thread: Production and Supply

WHAT IS A PRODUCT?

A product is often considered in a narrow sense as something tangible that can be described in terms of physical attributes, such as shape, dimension, components, form, colour, and so on. This is a misconception that has been extended to international marketing as well, because many people believe that only tangible products can be exported. A student of marketing, however, should realize that this definition of product is misleading since many products are intangible. Actually, intangible products are a significant part of the American export market.

For example, American movies are distributed worldwide, as are engineering services and business-consulting services. In the financial market, Japanese and European banks have been internationally active in providing financial assistance, often at handsome profits. Even when tangible products are involved, insurance services and shipping are needed to move the products into their markets.

In many situations, both tangible and intangible products must be combined to create a single, total product. Perhaps the best way to define a product is to describe it as *a bundle of utilities or satisfaction*. Warranty terms, for example, are a part of this bundle, and they may be adjusted as appropriate. Purchasers of Mercedes-Benz cars expect to acquire more than just the cars themselves. In hot and humid countries, there is no reason for a heater to be part of the automobile's product bundle.

In the USA, it is customary for automatic transmission to be included with other standard automobile equipment. One marketing implication that may be drawn is that a multinational marketer must look at a product as a total, complete offering. Consider the Beretta shotgun.

The shotgun itself is undoubtedly a fine product, quite capable of superbly performing its primary function. But Beretta also has a secondary function in Japan, where the Beretta brand is perceived as a superior status symbol. Not surprisingly, a Beretta can command $8000 for a shotgun, exclusive of the additional amount of a few thousand dollars for engraving. In this case, Beretta's secondary function conceivably overshadows its primary objective. Therefore, a complete product should be viewed as a satisfaction derived from the four Ps of marketing and not simply the physical product characteristics. Since a product can be bundled, it can also be unbundled. One problem with a bundled product is the increased cost associated with the extra benefits. With the increased cost, a higher price is inevitable.

Thus a proper marketing strategy, in some cases, is to unbundle a product instead so as to get rid of the frills and attract price-sensitive consumers. As an example, Serfin is a mid-tiered bank in Mexico, and is owned by Spain's Banco Santander Central Hispano. Serfin has launched Serfin Light, a new credit card that offers no points or air miles.

Instead, its key feature is an interest rate of 24 percent rather than 40 percent charged by the main competitors. The word "Light" is appropriate because Mexico is the world's largest consumer per capita of soft drinks, and Diet Coke is sold as Coke Light. The "light" concept has a significant meaning in Mexico. The success of Serfin Light prompted Banorte, the largest bank in northern Mexico, to change its slogan to "better than a light card, a strong card."

NEW PRODUCT DEVELOPMENT

There are six distinct steps in new product development. The *first step* is the *generation of new product ideas*. Such ideas can come from any number of sources. As in the case of Japan, already one out of five Japanese is age 65 or older, and the trend has adversely affected baby food.

From the peak of $252 million in 1999, sales of baby foods fell to $235 million in 2001. Searching for new sources of revenue, Japanese food companies were intrigued to learn that the same characteristics which make baby food appealing to babies also attracted old people. Thus food makers have come out with ready-to-eat treats: soft-boiled fish, bite-size shrimp meatballs, chop suey with tofu, and dozens of others.

These "Fun Meals" or "Food for Ages 0–100" only hint at the target demographic group without embarrassing older consumers. The *second step* involves the *screening of ideas*. Ideas must be acknowledged and reviewed to determine their feasibility.

To determine suitability, a new product concept may simply be presented to potential users, or an advertisement based on the product may be drawn and shown to focus groups to elicit candid reactions. As a rule, corporations usually have predetermined goals that a new product must meet. Kao Corporation, a major Japanese manufacturer of consumer goods, is guided by the following five principles of product development:

- A new product should be truly useful to society, not only now but also in the future,

- It should make use of Kao's own creative technology or skill,
- It should be superior to the new products of competitors, both from the standpoint of cost and performance,
- It should be able to stand exhaustive product tests at all stages before it is commercialized, and
- It should be capable of delivering its own message at every level of distribution.

The *third step* is *business analysis*, which is necessary to estimate product features, cost, demand, and profit. Xerox has small so-called product synthesis teams to test and weed out unsuitable ideas. Several competing teams of designers produce a prototype, and the winning model that meets preset goals then goes to the "product development" team.

The *fourth step* is *product development*, which involves lab and technical tests as well as manufacturing pilot models in small quantities. At this stage, the product is likely to be handmade or produced by existing machinery rather than by any new specialized equipment.

Ideally, engineers should receive direct feedback from customers and dealers. Goldstar Co., by letting its engineers out of the laboratories and into the market to see what Korean customers want, got an idea to make a refrigerator that can keep *kimchi* fresh and odourless for a long time. The refrigerator was an instant hit and enabled Goldstar to regain the top position which it lost to Samsung in South Korea in the late 1980s.

The *fifth step* involves *test marketing* to determine potential marketing problems and the optimal marketing mix. Anheuser Busch pulled Budweiser out of Germany after a six-month Berlin market test in 1981. Its Busch brand was another disappointment in France, where this type of beer did not yet correspond to French tastes.

Finally, assuming that things go well, the company is ready for *full-scale commercialization* by actually going through with full-scale production and marketing. It should be pointed out that not all of these six steps in new product development will be applicable to all products and countries. Test marketing, for example, may be irrelevant in countries where most major media are more national than local.

If the television medium has a nationwide coverage, it is not practical to limit a marketing campaign to one city or province for test marketing purposes. In any case, so many new products are tested and marketed each year. In Japan, because consumers constantly demand fresh, new products, some 700 to 800 drinks are launched annually.

To keep pace, Coca-Cola has built a product development center which allows it to cut launch time for new drinks from ninety days to a month, enabling it to release fifty new beverages a year. Unfortunately, it is easier for a new product to fail than to succeed. Naturally, so many things can go wrong. Therefore, it is just as crucial for a company to know when to retreat as when to launch a product. Coca-Cola's Ambasa Whitewater, a lacticbased drink, was removed from the market after eighteen months when sales started to decline.

MARKET SEGMENTATION

Market segmentation is a concept in economics and marketing. A market segment is a sub-set of a market made up of people or organizations with one or more characteristics that cause them to demand similar product and/or services based on qualities of those products such as price or function. A true market segment meets all of the following criteria: it is distinct from other segments, it is homogeneous within the segment; it responds similarly to a market stimulus, and it can be reached by a market intervention. The term is also used when consumers with identical product and/or service needs are divided up into groups so they can be charged different amounts for the services. The people in a given segment are supposed to be similar in terms of criteria by which they are segmented and different from other segments in terms of these criteria. These can be broadly viewed as 'positive' and 'negative' applications of the same idea, splitting up the market into smaller groups.

Examples:

- Gender
- Price
- Interests
- Location
- Religion
- Income
- Size of Household

While there may be theoretically 'ideal' market segments, in reality every organization engaged in a market will develop different ways of imagining market segments, and create Product differentiation strategies to exploit these segments. The market segmentation and corresponding product differentiation strategy can give a firm a temporary commercial advantage.

"POSITIVE" MARKET SEGMENTATION

Market segmenting is dividing the market into groups of individual markets with similar wants or needs that a company divides into distinct groups which have distinct needs, wants, behaviour or which might want different products and services. Broadly, markets can be divided according to a number of general criteria, such as by industry or public versus private. Although industrial market segmentation is quite different from consumer market segmentation, both have similar objectives. All of these methods of segmentation are merely proxies for true segments, which don't always fit into convenient demographic boundaries.

Consumer-based market segmentation can be performed on a *product specific* basis, to provide a close match between specific products and individuals.

However, a number of generic market segment systems also exist, *e.g.* the system provides a broad segmentation of the population of the United States based on the statistical analysis of household and geo-demographic data.

The process of segmentation is distinct from positioning. The overall intent is to identify groups of similar customers and potential customers; to prioritize the groups to address; to understand their behaviour; and to respond with appropriate marketing strategies that satisfy the different preferences of each chosen segment. Revenues are thus improved.

Improved segmentation can lead to significantly improved marketing effectiveness. Distinct segments can have different industry structures and thus have higher or lower attractiveness

Once a market segment has been identified and targeted, the segment is then subject to positioning. Positioning involves ascertaining how a product or a company is perceived in the minds of consumers.

This part of the segmentation process consists of drawing up a perceptual map, which highlights rival goods within one's industry according to perceived quality and price. After the perceptual map has been devised, a firm would consider the marketing communications mix best suited to the product in question.

USING SEGMENTATION IN CUSTOMER RETENTION

The basic approach to retention-based segmentation is that a company tags each of its active customers with 3 values:

- Tag 1: Is this customer at high risk of canceling the company's service? One of the most common indicators of high-risk customers is a drop off in usage of the company's service. For example, in the credit card industry this could be signaled through a customer's decline in spending on his or her card.
- Tag 2: Is this customer worth retaining? This determination boils down to whether the post-retention profit generated from the customer is predicted to be greater than the cost incurred to retain the customer. Managing Customers as Investments.
- Tag 3: What retention tactics should be used to retain this customer? For customers who are deemed "save-worthy", it's essential for the company to know which save tactics are most likely to be successful. Tactics commonly used range from providing "special" customer discounts to sending customers communications that reinforce the value proposition of the given service.

Process for tagging customers

The basic approach to tagging customers is to utilize historical retention data to make predictions about active customers regarding:

- Whether they are at high risk of canceling their service
- Whether they are profitable to retain
- What retention tactics are likely to be most effective

The idea is to match up active customers with customers from historic retention data who share similar attributes. Using the theory that "birds of a feather flock together", the approach is based on the assumption that active customers will have similar retention outcomes as those of their comparable predecessor.

PRICE DISCRIMINATION

Where a monopoly exists, the price of a product is likely to be higher than in a competitive market and the quantity sold less, generating monopoly profits for the seller. These profits can be increased further if the market can be segmented with different prices charged to different segments charging higher prices to those segments willing and able to pay more and charging less to those whose demand is price elastic.

The price discriminator might need to create rate fences that will prevent members of a higher price segment from purchasing at the prices available to members of a lower price segment. This behaviour is rational on the part of the monopolist, but is often seen by competition authorities as an abuse of a monopoly position, whether or not the monopoly itself is sanctioned. Examples of this exist in the transport industry where business class customers who can afford to pay may be charged prices many times higher than economy class customers for essentially the same service.

PRODUCT ADOPTION

In breaking into a foreign market, marketers should consider factors that influence product adoption. As explained by diffusion theory, at least six factors have a bearing on the adoption process: relative advantage, compatibility, trialability/divisibility, observability, complexity, and price. These factors are all perceptual and thus subjective in nature.

For a product to gain acceptance, it must demonstrate its *relative advantage* over existing alternatives. Products emphasizing cleanliness and sanitation may be unimportant in places where people are poor and struggle to get by one day at a time. Wool coats are not needed in a hot country, and products reducing static cling are useless in a humid country.

A sunscreen film attached to auto windshields to block out sunlight may be a necessity in countries with a tropical climate, but it has no such advantage in cold countries. Dishwashing machines do not market well in countries where manual labour is readily available and inexpensive. A product must also be *compatible* with local customs and habits. A freezer would not find a ready market in Asia, where people prefer fresh food. In Asia and such European countries as France and Italy, people like to sweep and mop floors daily, and thus there is no market for carpet or vacuum cleaners. Deodourants are deemed inappropriate in places where it is the custom for men to show their masculinity by having body odour. Dryers are unnecessary in countries where people prefer to hang their clothes outside for sunshine freshness. Kellogg's had difficulties selling Pop Tarts in Europe because many homes have no toaster. Unlike American women, European women do not shave their legs, and thus have no need for razors for that purpose.

The Japanese, not liking to have their lifestyles altered by technology, have skillfully applied technology to their traditional lifestyle. The electrical *kotatsu* is a traditional form of heater in Japan. New *kotatsu* are equipped with a temperature sensor and microcomputer to keep the interior temperature at a comfortable level. A new product should also be compatible with consumers' other belongings. If a new product requires a replacement of those other items that are still usable, product adoption becomes a costly proposition. A new product has an advantage if it is capable of being *divided* and *tested* in small trial quantities to determine its suitability and benefits.

This is a product's *trialability/divisibility* factor. Disposable diapers and blue jeans lend themselves to trialability rather well, but when a product is large, bulky, and expensive, consumers are much more apprehensive about making a purchase. Thus, washers, dryers, refrigerators, and automobiles are products that do not lend themselves well to trialability/ divisibility.

This factor explains one reason why foreign consumers do not readily purchase American automobiles, knowing that a mistake could ruin them financially. Many foreign consumers therefore prefer to purchase more familiar products, such as Japanese automobiles, that are less expensive and easier to service and whose parts are easier to replace. *Observation* of a product in public tends to encourage social acceptance and reinforcement, resulting in the product's being adopted more rapidly and with less resistance. If a product is used privately, other consumers cannot see it, and there is no prestige generated by its possession.

Blue jeans, quartz watches, and automobiles are used publicly and are highly observable products. Japanese men flip their ties so that labels show. Refrigerators, on the other hand, are privately consumed products, though owners of refrigerators in the Middle East and Asia may attempt to enhance observability by placing the refrigerator in the living room, where guests can easily see it. In any case, a distinctive and easily recognized logo is very useful. *Complexity* of a product or difficulty in understanding a product's qualities tends to slow down its market acceptance.

Perhaps this factor explains why ground coffee has had a difficult time in making headway to replace instant coffee in many countries. Likewise, 3M tried unsuccessfully in foreign markets to replace positive-acting printing plates with presensitized negative subtractive printing plates, which are very popular in the USA. It failed to convert foreign printers because the sales and technical service costs of changing printers' beliefs were far too expensive. Computers are also complex but have been gradually gaining more and more acceptance, perhaps in large part because manufacturers have made the machines simpler to operate. Ready-made software can also alleviate the necessity of learning computer languages, a timeconsuming process.

The first four variables are related positively to the adoption process. Like complexity, *price* is related negatively to product adoption. Prior to 1982, copiers were too big and expensive. Canon then introduced personal copiers with cartridges that customers could change. Its low price was so attractive to consumers that Canon easily dominated the market.

THEORY OF INTERNATIONAL PRODUCT LIFE CYCLE

The international product life cycle theory, developed and verified by economists to explain trade in a context of comparative advantage, describes the diffusion process of an innovation across national boundaries. The life cycle begins when a developed country, having a new product to satisfy consumer needs, wants to exploit its technological breakthrough by selling abroad. Other advanced nations soon start up their own production facilities, and before long less developed countries do the same.

Efficiency/comparative advantage shifts from developed countries to developing nations. Finally, advanced nations, no longer cost-effective, import products from their former customers.

The moral of this process could be that an advanced nation becomes a victim of its own creation. IPLC theory has the potential to be a valuable framework for marketing planning on a multinational basis. In this section, the IPLC is examined from the marketing perspective, and marketing implications for both innovators and initiators are discussed.

STAGES AND CHARACTERISTICS

There are five distinct stages in the IPLC. As the innovation moves through time, directions of all three curves change. Time is relative, because the time needed for a cycle to be completed varies from one kind of product to another. In addition, the time interval also varies from one stage to the next.

Stage 0 – Local innovation

Stage 0, depicted as time 0 on the left of the vertical importing/exporting axis, represents a regular and highly familiar product life cycle in operation within its original market. Innovations are most likely to occur in highly developed countries because consumers in such countries are affluent and have relatively unlimited wants. From the supply side, firms in advanced nations have both the technological know-how and abundant capital to develop new products. Many of the products found in the world's markets were originally created in the USA before being introduced and refined in other countries. In most instances, regardless of whether a product or not is intended for later export, an innovation is designed initially with an eye to capture the US market, the largest consumer nation.

Stage 1 – Overseas innovation

As soon as the new product is well developed, its original market well cultivated, and local demands adequately supplied, the innovating firm will look to overseas markets in order to expand its sales and profit. Thus this stage is known as a "pioneering" or "international introduction" stage. The technological gap is first noticed in other advanced nations because of their similar needs and high income levels.

Not surprisingly, English-speaking countries such as the United Kingdom, Canada, and Australia account for about half of the sales of US innovations when first introduced to overseas countries with similar cultures, and economic conditions are often perceived by exporters as posing less risk and thus are approached first before proceeding to less familiar territories.

Competition in this stage usually comes from US firms, since firms in other countries may not have much knowledge about the innovation. Production cost tends to be decreasing at this stage because by this time the innovating firm will normally have improved the production process.

Supported by overseas sales, aggregate production costs tend to decline further due to increased economies of scale. A low introductory price overseas is usually not necessary because of the technological breakthrough; a low price is not desirable due to the heavy and costly marketing effort needed in order to educate consumers in other countries about the new product.

In any case, as the product penetrates the market during this stage, there will be more exports from the USA and, correspondingly, an increase in imports by other developed countries.

Stage 2 – Maturity

Growing demand in advanced nations provides an impetus for firms there to commit themselves to starting local production, often with the help of their governments' protective measures to preserve infant industries.

Thus these firms can survive and thrive in spite of relative inefficiency. Development of competition does not mean that the initiating country's export level will immediately suffer.

The innovating firm's sales and export volumes are kept stable because LDCs are now beginning to generate a need for the product. Introduction of the product in LDCs helps offset any reduction in export sales to advanced countries.

Stage 3 – Worldwide imitation

This stage means tough times for the innovating nation because of its continuous decline in exports. There is no more new demand anywhere to cultivate. The decline will inevitably affect the US innovating firm's economies of scale, and its production costs thus begin to rise again. Consequently, firms in other advanced nations use their lower prices to gain more consumer acceptance abroad at the expense of the US firm.

As the product becomes more and more widely disseminated, imitation picks up at a faster pace. Towards the end of this stage, US export dwindles almost to nothing, and any US production still remaining is basically for local consumption. The US automobile industry is a good example of this phenomenon. There are about thirty different companies selling cars in the USA, with several on the rise. Of these, only two are US firms, with the rest being from Western Europe, Japan, South Korea, and others.

Stage 4 – Reversal

Not only must all good things end, but misfortune frequently accompanies the end of a favourable situation. The major functional characteristics of this stage are *product standardization and comparative disadvantage*. This innovating country's comparative advantage has disappeared, and what is left is comparative disadvantage.

This disadvantage is brought about because the product is no longer capital-intensive or technology-intensive but instead has become labour-intensive – a strong advantage possessed by LDCs. Thus LDCs – the last imitators – establish sufficient productive facilities to satisfy their own domestic needs as well as to produce for the biggest market in the world, the USA.

US firms are now undersold in their own country. Black-and-white TV sets, for example, are no longer manufactured in the USA because many Asian firms can produce them much less expensively than any US firm. Likewise, the USA hardly produces colour TV sets either. Consumers' price sensitivity exacerbates this problem for the initiating country.

VALIDITY OF THE IPLC

Several products have conformed to the characteristics described by the IPLC. The production of semiconductors started in the USA before diffusing to the United Kingdom, France, Germany, and Japan. Production facilities are now set up in Hong Kong and Taiwan, as well as in other Asian countries. Similarly, at one time, the USA used to be an exporter of typewriters, adding machines, and cash registers.

However, with the passage of time, these simple machines are now being imported, while US firms export only the sophisticated, electronic versions of such machines. Other products that have gone through a complete international life cycle are synthetic fibres, petrochemicals, leather goods, rubber products, and paper. The electronics sector, a positive contributor to the trade balance of the USA for a long time, turned negative for the first time ever in 1984 with a massive $6.8 billion deficit.

A deficit also occurred at the same time for communications equipment, following the trend set by semiconductors in 1982. The IPLC is probably more applicable for products related through an emerging technology. These newly emerging products are likely to provide functional utility rather than aesthetic values. Furthermore, these products likely satisfy basic needs that are universally common in most parts of the world. Washers, for example, are much more likely to fit this theory than are dryers. Dishwashing machines are not useful in countries where labour is plentiful and cheap, and the diffusion of this kind of innovation as described in IPLC is not likely to occur.

MARKETING STRATEGIES

For those advanced economies' industries in the worldwide imitation stage or the maturity stage, things are likely to get worse rather than better. The prospect, though bleak, can be favourably influenced. What is crucial is for firms in the advanced economies to understand the implications of the IPLC so that they can adjust marketing strategies accordingly.

Product policy

The IPLC emphasizes the importance of cost advantage. It would be very difficult for firms in advanced economies to match labour costs in low-wage nations since costs are only 0.5 cent in China. Still, the innovating firm must keep its product cost competitive. One way is to cut labour costs through automation and robotics.

IBM converted its Lexington plant into one of the most automated plants in the world. Japanese VCR manufacturers are counting on automation to help them meet the challenge of South Korea. Another way to reduce production costs is to eliminate unnecessary options, since such options increase inefficiency and complexity.

This strategy may be crucial for simple products or for those at the low end of the price scale. In such cases, it is desirable to offer a standardized product with a standard package of features or options included. To keep costs rising at a minimum, an initiating firm may use local manufacturing in other countries as an entry strategy.

The company can not only minimize transportation costs and entry barriers but also indirectly slow down potential local competition starting up manufacturing facilities. Another benefit is that those countries can eventually become a springboard for the company to market its product throughout that geographic region. In fact, sourcing should allow the innovator to hold down labour costs at home and abroad and retain the original market as well. Manufacturers should examine the traditional vertical structure in which they make all or most components and parts themselves because in many instances outsourcing may prove to be more costeffective. Outsourcing is the practice of buying parts or whole products from other manufacturers while allowing a buyer to maintain its own brand name. For example, Ford Festiva is made by Kia Motors, Mitsubishi Precis by Hyundai, Pontiac Lemans by Daewoo, and GM Sprint by Suzuki.

A modification of outsourcing involves producing various components or having them produced under contract in different countries. That way, a firm takes advantage of the most abundant factor of production in each country before assembling components into final products for worldwide distribution. Solectron and Flextronics are examples of contract manufacturers that do manufacturing for many well-known brands. Solectron, a contract electronics maker, makes components and finished products for electronics companies, and its customers include Cisco Systems, HP, and Ericsson.

A recent deal involves Solectron making optical networking equipment worth as much as $2 billion for Lucent Technologies for three years. Xerox has a five-year contract that transfers about half of Xerox's manufacturing operations to Flextronics and represents more than $1 billion in annual manufacturing costs. Flextronics, based in Singapore, is a $12 billion global electronics manufacturing services company, manufacturing Xerox office equipment and components at a modest premium over book value. These copiers and printers are used worldwide.

Once in the maturity stage, the innovator's comparative advantage is gone, and the firm should switch from producing simple versions to producing sophisticated models or new technologies in order to remove itself from cut-throat competition. Japanese VCR makers used to make 99 percent of the machines sold in the USA and 75 percent of all machines sold worldwide, but they still cannot compete with low-wage Korean newcomers strictly on price, because labour content in VCRs is substantial.

To retain their market share, the Japanese rely on new technology, such as 8-mm camcorders. For a relatively high-tech product, an innovator may find it advantageous to get its product system to become the industry's standard, even if it means lending a helping hand to competitors through the licensing of product knowledge. Otherwise, there is always a danger that competitors will persevere in inventing an incompatible and superior system.

A discussion of product adoption above should make it clear that several competing and incompatible systems serve only to confuse potential adopters who must acquire more information and who are uncertain as to which system will survive over the long term. The worst scenario for an innovating firm is when another system supplants the innovators' product altogether to become the industry's standard. Sony's strategic blunder in guarding its Betamax video system is a good case to study. Matsushita and Victor Co. took the world leadership position away from Sony by being more liberal in licensing its VHS to their competitors.

Philips and Grundig did not introduce their Video 2000 system until VHS was just about to become the industry's standard in Europe and the world. By that time, despite price cutting, it was too late for Video 2000 to attract other manufacturers and consumers. The problem was so bad that Philips' own North American subsidiary refused to buy its parent's system. Ironically, Sony itself had to start making VHS-format machines in 1988. In 2002, Sony ended its bitter experience by discontinuing the Betamax machines.

A more recent case of competing technologies and strategic alliances involves the digital videodisc which can also store audio and computer data and software. Toshiba's system competed with the Multimedia Compact Disc system offered by Sony and Philips. In addition, Toshiba aggressively courted movie makers while offering "open" licensing to other electronics companies. As a result, such manufacturing giants as Matsushita, Thomson, and Pioneer chose to ally with Toshiba.

In the end, Sony and Philips had to come to a compromise by adopting a single format that was closer to Toshiba's system than theirs. Matsushita and Sony recently faced off again as both tried to make their products the industry standard so as to control the market of related digital consumer products. The battle involved flash memory cards which are used to record data on digital cameras, music players, and next-generation mobile phones and computers. Matsushita, playing catch-up, has the support of nearly ninety manufacturers worldwide, while Sony has fifty-eight. In addition, Matsushita's DVD audio player is not compatible with Sony's SACD, and its first recordable DVD player uses DVD-RAM that is not compatible with the DVD-RW format led by Pioneer and supported by Sony.

Pricing policy

Initially, an innovating firm can afford to behave as a monopolist, charging a premium price for its innovation. But this price must be adjusted downward in the second and third stages of IPLC to discourage potential newcomers and to maintain market share. Anticipating a Korean challenge, Japanese VCR makers cut their prices in the USA by 25 percent and were able to slow down retailers' and consumers' acceptance of Korean brands. IBM, in comparison, was slow in reducing prices for its PC models.

The error in judgement was the result of a belief that the IBM PC was too complex for Asian imitators. This proved to be a costly error because the basic PC hardly changed for several years. Before long, the product became nothing but an easily copied, standardized commodity suitable for intensive distribution – the kind that Asian companies thrive on. Commodity pricing now dominates the market. In the end, IBM stopped manufacturing desktop PCs in most parts of the world. Instead, its PCs are now made by Sanmina- SCI, a contract manufacturer. In the last stage of the IPLC, it is not practical for the innovating firm to maintain low price because of competitors' cost advantage. However, the firm's above-the-market price is feasible only if it is accompanied by top-quality or sophisticated products. A high standard of excellence should partially insulate the firm's product from direct price competition. US car makers failed in this area – high prices are not matched by consumer perception of superior quality.

Promotion policy

Promotion and pricing in the IPLC are closely related. The innovating firm's initial competitive edge is its unique product, which allows it to command a premium price. To maintain this price in the face of subsequent challenges from imitators, uniqueness can be retained only in the form of superior quality, style, or services. The innovating marketer must plan for a non-price promotional policy at the outset of a product diffusion. Timken is able to compete effectively against the Japanese by offering more services and meeting customers' needs at all times.

For instance, it offers technological support by sending engineers to help customers design bearings in gearboxes. One implication that may be drawn is that a new product should be promoted as a premium product with a high-quality image. By starting out with a high-quality reputation, the innovating company can trade down later with a simpler version of the product while still holding on to the high-priced, most profitable segment of the market.

One thing the company must never do is to allow its product to become a commodity item with prices as the only buying motive, since such a product can be easily duplicated by other firms. Aprica has been very successful in creating a status symbol for its stroller by using top artists and designers to create a product for mothers who are more concerned with style than price. The stroller is promoted as "anatomically correct" for babies to avoid hip dislocation, and the company uses the snob appeal and comfort to distinguish its brand from those of Taiwanese and Korean imitators.

Therefore, product differentiation, not price, is most important for insulating a company from the crowded, low-profit market segment. A product can be so standardized that it can be easily duplicated, but image is a very different proposition.

A more recent example is a treatment for impotence. Viagra was the first to hit the market, generating an incredible amount of discussion worldwide. It is important for Viagra to create brand awareness and preference. Within a few short years, a number of rivals entered the market with improved products. Bayer and GSK, using vardenafil, markets Levitra which works regardless of the cause of a patient's impotence – depression, heart disease, high blood pressure, diabetes, and so on. Eli Lilly's Cialis claims to work for up to thirtysix hours or seven times as long as Viagra, thus having an advantage in terms of picking the right moment.

Place policy

A strong dealer network can provide the US innovating firm with a good defensive strategy. Because of its near-monopoly situation at the beginning, the firm is in a good position to be able to select only the most qualified agents/distributors, and the distribution network should be expanded further as the product becomes more diffused. Caterpillar's network of loyal dealers caused difficulty for Komatsu to line up its own dealers in the USA.

In an ironic case, GM's old policy of limiting its dealers from carrying several GM brands inadvertently encouraged those dealers to start carrying imports. A firm must also watch closely for the development of any new alternative channel that may threaten the existing channel. When it is too late or futile to keep an enemy out, the enemy should be invited in. US firms – manufacturers as well as agents/distributors – can survive by becoming agents for their former competitors.

This tactic involves providing a distribution network and marketing expertise at a profit to competitors who in all likelihood would welcome an easier entry into the marketplace. American car makers and their dealers seem to have accepted the reality of the marketplace and have become partners with their Japanese and foreign competitors, as evidenced by General Motors's ventures with Toyota, Suzuki, and Isuzu, American Motors' with France's Renault, Chrysler's with Mitsubishi and Maserati, and Ford's with Mazda and the Korean Kia Motors. Once a product is in the final stage of its life cycle, the innovating firm should strive to become a specialist, not a generalist, by concentrating its efforts in carefully selected market segments, where it can distinguish itself from foreign competitors.

To achieve distinction, US firms should either add product features or offer more services. For the alert firm, there are early warning signals that may be used to determine whether the time has come to adopt this strategy. One signal is that the product becomes so standardized that it can be manufactured in many LDCs.

Another warning signal is a decline in the US exports owing to the loss of narrowing of the US technological lead. By that time, certain forms of market segmentation and product differentiation are highly recommended. As in the case of consumer electronics, such great American brands as Marantz and Scott were once synonymous with good sound and top quality but have since been bought up or driven out of business by Japanese and European manufacturers. American firms continue to dominate the segment of high-end stereo equipment where top systems may cost up to $100,000. American firms are successful because of the precision required and because production runs are short and usually done by hand.

PRODUCT STANDARDIZATION VS. PRODUCT ADAPTATION

Product standardization means that a product designed originally for a local market is exported to other countries with virtually no change, except perhaps for the translation of words and other cosmetic changes. Revlon, for example, used to ship successful products abroad without changes in product formulation, packing, and advertising. There are advantages and disadvantages to both standardization and individualization.

ARGUMENTS FOR STANDARDIZATION

The strength of standardization in the production and distribution of products and services is its *simplicity* and *cost*. It is an easy process for executives to understand and implement, and it is also cost effective. If cost is the only factor being considered, then standardization is clearly a logical choice because economies of scale can operate to reduce production costs. Yet minimizing production costs does not necessarily mean that profit increases will follow.

Simplicity is not always beneficial, and costs are often confused with profits. Cost reductions do not automatically lead to profit improvements, and in fact the reverse may apply. By trying to control production costs through standardization, the product involved may become unsuitable for alternative markets.

The result may be that demand abroad will decline, which leads to profit reduction. In some situations, cost control can be achieved but at the expense of overall profit. It is therefore prudent to remember that cost should not be overemphasized. The main marketing goal is to maximize profit, and production-cost reductions should be considered as a secondary objective. The two objectives are not always convergent.

When appropriate, standardization is a good approach. For example, when a *consistent company or product image* is needed, product uniformity is required. The worldwide success of McDonald's is based on consistent product quality and services. Hamburger meat, buns, and fruit pies must meet strict specifications. This obsession with product quality necessitates the costly export of french fries from Canada to European franchises because the required kind of potato is not grown in Europe.

In 1982, a Paris licensee was barred through a court order from using McDonald's trademarks and other trade processes because the licensee's twelve Paris eateries did not meet the required specifications. Some products by their very nature are not or cannot be easily modified. *Musical recordings and works of art* are examples of products that are difficult to differentiate, as are books and motion pictures. When this is the case, the product must rise and fall according to its own merit. Whether such products will be successful in diverse markets is not easy to predict. Films that do well in the USA may do poorly in Japan. On the other hand, movies that were not box-office hits in the USA have turned out to be money makers in France.

Yet in the case of Ricky Martin, his worldwide fame has to do in part with his France '98 World Cup theme song – in Spanish, English, and French. With regard to high-technology products, both users and manufacturers may find it desirable to reduce confusion and promote compatibility by introducing *industry specifications* that make standardization possible.

A condition that may support the production and distribution of standardized products exists when certain products can be associated with particular *cultural universals*; that is, when consumers from different countries share similar need characteristics and therefore want essentially identical products. Watches are used to keep time around the world and thus can be standardized. The diamond is another example.

Levi Strauss' attempt to penetrate the European market with lighter-weight jeans failed because European consumers preferred the standard heavy-duty American type. Another study also found that industrial managers and managers of consumer goods regarded certain marketing-related factors differently, thus implying that product standardization or customization depends in part on the type of product.

Furthermore, respondents consistently regarded competitive environment as the most important variable affecting the extent of marketing standardization. While market conditions tend to support localization, Western companies tend to favour standardization, but there is some logic behind the practice. Most Central and Eastern European countries are too small to pay off for customization. In addition, within a decade, these markets may converge to Western Europe market structures and rules.

ARGUMENTS FOR ADAPTATION

There is nothing wrong with standardized products if consumers prefer those products. In many situations, domestic consumers may desire a particular design of a product produced for the American market. However, when the product design is placed in foreign markets, foreign buyers are forced either to purchase that product from the manufacturer or not purchase anything at all. This manner of conducting business overseas is known as the "big-car" and "left-hand-drive" syndromes.

Both describe US firms' reluctance and/or unwillingness to modify their products to suit their customers' needs. According to the big-car syndrome, US marketers assume that products designed for Americans are superior and will be preferred by foreign consumers. US car makers believe that the American desire for big cars means that only big cars should be exported to overseas markets. The left-hand-drive syndrome is a corollary to the big-car syndrome. Americans drive on the right-hand side of the road, with the steering wheel on the left side of the automobile. But many Asian and European countries have traffic laws requiring drivers to drive on the left side of the road, and cars with steering wheel on the left present a serious safety hazard. Yet exported US cars are the same left-hand-drive models as are sold in the USA for the right-hand traffic patterns.

According to the excuse used by US car makers, a small sales volume abroad does not justify converting exported cars to right-hand steering: as GM's president explained, "There's a certain status in having a left-hand steer in Japan." This is one good reason why American automobile sales abroad have been disappointing.

A half-hearted effort can only result in a half-hearted performance. American exporters have failed time and time again to realize that when in Rome one should do as the Romans do. Japanese car makers have not shown the same kind of indifference to market needs; Japanese firms have always adapted their automobiles to American driving customs.

Those American firms that have understood the need for product modification have done well even in Japan. Du Pont has customized its manufacturing and marketing for the Japanese market, and its design units work with Japanese customers to design parts to their specifications. Sprite became the best selling clear soft drink in Japan after being reformulated – the lime taste was taken out because Japanese were found to prefer a purer lemon flavor. Yamazaki-Nabisco's Ritz crackers sold in Japan are less salty than the Ritz crackers sold in America; similarly, Chips Ahoy are less sweet than versions sold elsewhere.

Responding to the Japanese demand for quality, Ajinomoto-General Foods used a better grade of bean for its Maxim instant coffee. Firms must choose the time when a product is to be modified to better suit its market. According to the Conference Board, important factors for product modification mentioned by more than 70 percent of firms surveyed are long-term profitability, long-term market potential, product–market fit, short-term profitability, cost of altering or adapting, desire for consistency and short-term market potential. These factors apply to consumer non-durable and durable products as well as to industrial products.

Product adaptation is necessary under several conditions. Some are mandatory, whereas others are optional. In addition, firm characteristics and environmental characteristics have a significant impact on a firm's overall performance and marketing mix strategy. Mandatory product modification The mandatory factors affecting product modifications include the following:

government's mandatory standards, electrical current standards, measurement standards, and product standards and systems. The most important factor that makes modification mandatory is government regulation. To gain entry into a foreign market, certain requirements must be satisfied. Regulations are usually specified and explained when a potential customer requests a price quotation on a product to be imported. Avon shampoos had to be reformulated to remove the formaldehyde preservative, which is a violation of regulations in several Asian countries.

Food products are usually heavily regulated. Added vitamins in margarine, forbidden in Italy, are compulsory in the United Kingdom and Holland. In the case of processed cheese, the incorporation of a mold inhibitor may be fully allowed, allowed up to the permissible level, or forbidden altogether. Frequently, products must be modified to compensate for differences in electrical current standards. In many countries there may even be variations in electrical standards within the country.

The different electrical standards abroad can easily harm products designed for use in the USA, and such improper use can be a serious safety hazard for users as well. Stereo receivers and TV sets manufactured for the US 110- to 120-volt mode will be severely damaged if used in markets where the voltages are twice as high.

Therefore, products must be adapted to higher voltages. When there is no voltage problem, a product's operating efficiency may be impaired if the product is operated in the wrong electrical frequency. Alarm clocks, tape-recorders, and turntables designed for the US 60-Hz system will run more slowly in countries where the frequency is 50 Hz. To solve this problem, marketers may have to substitute a special motor or arrange for a different drive ratio to achieve the desirable operating RPM or service level.

As with electrical standards, measurement systems can also vary from country to country. Although the USA has adopted the English system of measurement, most countries employ the metric system, and product quantity should or must be expressed in metric units. Starting in 2010, the EU countries will no longer accept non-metric products for sale. Many countries even go so far as to prohibit the sale of measuring devices with both metric and English markings.

One New England company was ordered to stop selling its laboratory glassware in France because the markings were not exclusively metric. In 1982, in order to save $2 million, the US Congress abolished the US Metric Board as well as a voluntary programme for conversion to the metric system. That decision was short sighted.

Metric demands adversely affect US firms' competitiveness because many American firms do not offer metric products. A Middle-Eastern firm, for example, was unable to find an American producer that sold pipe with metric threads for oil machinery. A European firm had to rewire all imported electrical appliances because the US standard wire diameters did not meet national standards. It is difficult to find a US firm that cuts lumber to metric dimensions. Fortunately, the new trade act now requires the US government and industry to use metric units in documentation of exports and imports as prescribed by the International Convention on the Harmonized Commodity Description and Coding System. The Harmonized System is designed to standardize commodity classification for all major trading nations. The International System Units is the official measurement system of the Harmonized System. Very few countries still cling to the obsolete non-metric systems. Among them are the USA, Burma, Brunei, and Liberia. Robert Heller of the Federal Reserve Board of Governors made the following comment:

Only Yemen and India have as low an export-to- GDP ratio as the USA. Would it come as a surprise to you to know that the US and Yemen share something else in common? They are the only two countries in the world that have not yet gone metric! If an American manufacturer has to retool first in order to sell his wares abroad, his incentive to do so is considerably reduced, and it makes his first step into export markets all that much more expensive.

Some products must be modified because of different operating systems adopted by various countries. Television systems provide a good example. There are three different TV operating systems used in different parts of the world: the American NTSC, the French SECAM and the German PAL. In 1941, the USA became the first country to set the national standards for TV broadcasting, adopting 525 scanning lines per frame. Most other nations later decided to adopt 625 lines for a sharper image. In most cases, a TV set designed for one broadcast system cannot receive signals broadcast through a different operating system. When differences in product operating systems exist, a company unwilling to change its products must limit the number of countries it can enter, unless proper modification is undertaken for other market requirements.

Optional product modification

The conditions dictating product modification mentioned so far are mandatory in the sense that without adaptation a product either cannot enter a market or is unable to perform its function there. Such mandatory standards make the adaptation decision easy: a marketer must either comply or remain out of the market.

Italy's Piaggio withdrew its Vespa scooters from the US market in 1983, choosing not to meet US pollution control standards for its few exports. A more complex and difficult decision is optional modification, which is based on the international marketer's discretion in taking action. Nescafé in Switzerland, for instance, tastes quite different from the same brand sold just a short distance across the French border. One condition that may make optional modification attractive is related to physical distribution, and this involves the facilitation of product transportation at the lowest cost. Since freight charges are assessed on either a weight or volume basis, the carrier may charge on the basis of which ever is more profitable. The marketer may be able to reduce delivery costs if the products are assembled and then shipped. Many countries also have narrow roads, doorways, stairways, or elevators that can cause transit problems when products are large or are shipped assembled. Therefore, a slight product modification may greatly facilitate product movement. Another determinant for optional adaptation involves local use conditions, including climatic conditions.

The hot/cold, humid/dry conditions may affect product durability or performance. Avon modified its Candid moist lipstick line for a hot, humid climate. Certain changes may be required in gasoline formulations. If the heat is intense, gasoline requires a higher flashpoint to avoid vapor locks and engine stalling. In Brazil, automobiles are designed to run on low-quality gas, to withstand the country's rough dusty roads, and to weather its sizzling temperatures.

As a result, these automobiles are attractive to customers in LDCs, especially when the automobiles are also durable and simple to maintain. American automobiles can experience difficulties in these markets, where people tend to overload their cars and trucks and do not perform regular maintenance, not to mention the unavailability of lead-free gasoline. Another local use condition that can necessitate product change is space constraint. Sears' refrigerators were redesigned to be smaller in dimensions without sacrificing the original capacity, so that they could fit into the compact Japanese home. Philips, similarly, had to reduce the size of its coffee maker.

In contrast, US mills, for many years, resisted cutting plywood according to Japanese specifications, even though they were told repeatedly that the standard Japanese plywood dimensions were 3 × 6 ft – not the US standard of 4 × 8 ft. In a related case, Japanese-style homes have exposed wood beams, but US forest-products firms traditionally allow 2 × 4 studs to be dirty or slightly warped, since in the USA these studs will almost always be covered over with wallboard.

The firms have refused to understand that wood grain and quality are important to the Japanese because an exposed post is part of the furniture. Furniture is not easily exported because it has two inherent problems: size/weight and different ways furniture is designed and used in other cultures. Some foreign manufacturers are still able to be successful in Japan, especially those who are willing to reduce the size of some products and make necessary modifications so as to appeal to Japanese consumers. Consumer demographics as related to physical appearance can also affect how products are used and how suitable those products are. Habitat Mothercare PLC found out that its British products were not consistent with American customs and sizes. Its comforters were not long enough to fit American beds, and its tumblers could not hold enough ice. Philips downsized its shavers to fit the smaller Japanese hand. One US brassière company did well initially in Germany yet failed to win repeat purchases.

The problem was that German women have a tendency not to try on merchandise in the store and thus did not find out until later that the product was ill-fitting because of measurement variations between American and German brassières.

Furthermore, German customers do not usually return a product for refund or adjustment. Even a doll may have to be modified to better resemble the physical appearance of local people. The Barbie doll, though available in Japan for decades, became popular only after Mattel allowed Takara to reconstruct the product. Out of sixty countries, Japan is the only market where the product is modified. Barbie's Western-style features are modified in several ways: her blue eyes become brown, her vividly blonde hair is darkened, and her bosom size is reduced.

Local use conditions include users' habits. Since the Japanese prefer to work with pencils – a big difference from the typed business correspondence common in the USA – copiers require special characteristics that allow the copying of light pencil lines. Microsoft's plant in Ireland was charged with the task of localizing Windows 95 into more than fifty languages. IBM, likewise, localized its OS/2 Warp system. It took four months for IBM to translate it for the Czech Republic.

The words on the screen had to be translated first from Czech to English. Later, the programme was adapted to the Czech operating system. However, on the first try, the OS/2 could not accept any Windows applications due to the differences in the Czech system. Another month of adjustment was necessary before the Czech version was ready.

The Polish, Hungarian, and Russian language versions were also made available. Finally, other environmental characteristics related to use conditions should be examined. Examples are endless. Detergents should be reformulated to fit local water conditions.

IBM had to come up with a completely new design so that its machine could include Japanese word-processing capability. Kodak made some changes in its graphic arts products for Japanese professionals, most of whom have no dark-rooms and have to work in different light environments. Price may often influence a product's success or failure in the marketplace. This factor becomes even more crucial abroad because US products tend to be expensive, but foreign consumers' incomes tend to be at lower levels than Americans' incomes. Frequently, the higher quality of American products cannot overcome the price disadvantage found in foreign markets. To solve this problem, American companies can reduce the contents of the product or remove any non-essential parts or do both. Foreign consumers are generally not convenience oriented, and an elaborate product can be simplified by removing any "frills" that may drive up the price unnecessarily.

This approach is used by General Motors in manufacturing and selling the so-called Basic Transportation Vehicle in less-industrialized nations. One reason that international marketers often voluntarily modify their products in individual markets is their desire to maximize profit by limiting product movement across national borders.

The rationale for this desire to discourage gray marketing is that some countries have price controls and other laws that restrict profits and prices. When other nearby countries have no such laws, marketers are encouraged to move products into those nearby countries where a higher price may be charged. A problem can arise in which local firms in countries where product prices are high are bypassed by marketers who buy directly from firms handling such products in countries where prices are low.

In many cases, due to antitrust laws, international marketers who wish to maintain certain market prices cannot ban this kind of product movement by threatening to cut off supply from those firms re-exporting products to high-priced countries. Johnson and Johnson, for example, was fined $300,000 by the EU for explicitly preventing British wholesalers and pharmacists from re-exporting Gravindex pregnancy tests to Germany, where the kits cost almost twice as much. In spite of authorities' efforts to prevent companies from keeping lower-priced goods out of higher-priced countries, marketers may do so anyway as long as they do not get caught.

Some manufacturers try to hinder these practices by deliberately varying packaging, package coding, product characteristics, colouring, and even brand names in order to spot violators or to confuse consumers in markets where products have moved across borders. Perhaps the most arbitrary yet most important reason for product change abroad is because of historical preference, or local customs and culture.

Product size, colour, speed, grade, and source may have to be redesigned in order to accommodate local preference. Kodak altered its film to cater to a Japanese idea of attractive skin tones. Kraft's Philadelphia Cream Cheese tastes different in the USA, Great Britain, Germany, and Canada.

In Asia, Foremost sells chocolate and strawberry milk instead of lowfat and skimmed milk. Asians and Europeans by tradition prefer to shop on a daily basis, and thus they desire smaller refrigerators in order to reduce cost and electrical consumption. When products clash with a culture, the likely loser is the product, not the culture. Strong religious beliefs make countries of the Middle East insist on halaled chickens. In soup-conscious Brazil, Campbell soups did not take off because homemakers there have strong cultural traditions of a home-maker's role, and serving Campbell soup to their families would be a soup served not of their own making. As a result, these home makers prefer dehydrated products manufactured by Knorr and Maggi, used as a soup starter to which the homemaker can add her own flavors and ingredients. Campbell soups are usually purchased to be set aside for an emergency, such as if the family arrives home late.

Product changes are not necessarily related to functional attributes such as durability, quality, operation method, maintenance, and other engineering aspects. Frequently, aesthetic or secondary qualities must also be taken into account. There are instances in which minor, cosmetic changes have significantly increased sales.

Therefore, functional and aesthetic changes should both be considered in regard to how they affect the total, complete product. One company that incorporates multiple features of product modification in appealing to local tastes is Pillsbury.

In marketing its Totino's line of pizzas in Japan, Pillsbury found it wise to make several mandatory and optional changes in its product. Japanese food standards ban many preservatives and dyes.

The ban often necessitates an extensive redesign of a product just to get it into the Japanese market.

Totino's pizzas are basically a "belly stuffer" in the USA, a confirmation of many foreigners' perceptions that Americans have pedestrian tastes in food. But the Japanese "eat" with their eyes, too – all foods have an aesthetic dimension.

They perceive American foods as being too sweet, too large, and too spicy, making it necessary to alter the ingredients to suit the Japanese palate. Furthermore, the pizza size had to be reduced from the US 12-ounce size to the Japanese 6.5-ounce size to fit into smaller Japanese ovens.

In effect, Totino's frozen pizza and packaging were completely redesigned for the Japanese market, and Pillsbury's success confirms that its efforts were worthwhile. In conclusion, marketers should not waste time resisting product modification.

The reluctance to change a product may be the result of an insensitivity to cultural differences in foreign markets. Whatever the reason for this reluctance, there is no question that it is counter-productive in international marketing. Product adaptation should rarely become an important issue to the marketer. A good marketer compares the incremental profits against the incremental costs associated with product adaptation.

If the incremental profit is greater than the associated incremental cost, then the product should be adjusted – without question. In making this comparison, marketers should primarily use only future earnings and costs.

A MOVE TOWARDS WORLD PRODUCT: INTERNATIONAL OR NATIONAL PRODUCT?

Product standardization and modification may give the impression that a marketer must choose between these two processes and that one approach is better than the other. In many instances, a compromise between the two is more practical and far superior than selecting either procedure exclusively. Black and Decker has stopped customizing products for every country in favour of a select few global products that may be sold everywhere.

Such US publishers as Prentice-Hall and Harper have also adopted the "world book" concept, which makes it possible for an English-language world book to have world copyright. Publishers change, if necessary, only the title-page, cover, and/or jacket. World product and standardized product may sometimes be confused with each other.

A world product is a product designed for the international market. In comparison, a standardized product is a product developed for one national market and then exported with no change to international markets. Zenith and RCA TV sets are standardized products, whereas a German subsidiary of ITT makes a world product by producing a "world chassis" for its TV sets. This world chassis allows assemblage of TV sets for all three colour TV systems of the world without changing all circuitry on the various modules.

A move by a company towards a world product is a logical and healthy move. If a company has to adapt its product for each market this can be a very expensive proposition, but without the necessary adaptation a product may not sell at all. Committing to the design of a world product can provide the solution to these two major concerns faced by most firms dealing in the international marketplace.

GE, for example, produces a numerical control system suitable in both metric and English measures. In addition, it has designed machines to operate under the wide differences in voltage among the different European countries. GE refrigerators are built in such a way that they can be used regardless of whether the frequency is 50 Hz or 60 Hz.

This emerging trend towards world products is also attractive for items with international appeal or for those items purchased by international travelers. Electric shavers made by Norelco and portable stereo radios made by Sony and Crown are produced having a universal-voltage feature. One might question whether a world product would be more expensive than a national or local product, since the world product may need multipurpose parts.

Actually, the world product should result in greater saving for two reasons. First, costly downtime in production is not needed to adjust or convert equipment to produce different national versions. Second, a world product greatly simplifies inventory control because only one universal part, rather than many individual parts, has to be stocked. A world product may also be able to lower certain production costs by anticipating necessary local adaptation and thus being adaptation-ready. As an example, the Japanese government requires thirtytwo changes on most US-built cars, including: replacing headlamps that, because of the left-hand drive, dip in the wrong direction; changing "sharpedged" door handles; replacing outside rear view mirrors; and filling in the gap between the body and the rear bumper to prevent catching the sleeves of kimono-clad women. Honda is able to sell its US made cars in Japan at relatively low prices because it produces the car ready for sale in Japan.

Since cars manufactured by GM, Ford, and Chrysler are built for the American market, they must undergo expensive alterations to meet Japanese regulations. The American car makers have taken some steps to remedy the problem. It must be pointed out that a world product has some inherent problems as well. As illustrated by Ford Escort, the car was designed in Europe as Ford's world car.

The company's American executives, however, proceeded to thoroughly redesign it for the US market. Similarly, design choices are often a source of conflict between Toyota's Japanese headquarters and its US subsidiary. Even interior colour schemes, while minor, are also in dispute. Therefore, corporate commitment is a necessity.

There must be mechanisms to take care of the conflicting views of executives working in the different countries. The Ford 2000 project, relying on centralization, uses vehicle centers to better use common parts and engineering ideas. However, due to differences in gasoline prices, consumer income, and tastes, the same vehicle does not have the same appeal in many parts of the world. While Americans are crazy about Ford's trucks and sports-utility vehicles, Europeans like smaller, more fuel-efficient passenger cars. While Europeans prefer flashier interiors, their American counterparts want wider seats while toning down the interiors.

As a product of compromise, a world product may have to be bland enough to partially please everyone while not really pleasing anyone; that is, it must satisfy the lowest common denominator of taste in different markets. Ford's Mondeo has done well in Europe, but American consumers have found the back seat of the American versions to be too tight. Likewise, GM's 1997 front-drive minivan is just right for the Europeans but a little too small for the Americans. As far as the automobile industry is concerned, a world car has another problem: it has to meet the world's toughest environmental and safety rules, thus increasing costs.

The trend towards an international or world product and away from a national product will continue as MNCs become more aware of the significance of world marketing. The willingness of several companies to consider designing a universal product for the world market is indeed a good indicator that this trend will continue. Consider the case of Vaillant. This German boiler company is Europe's biggest maker of central heating boilers. The boiler market does not accommodate the pan-European plans well. Due to a huge variation in customer tastes and building standards, a company has to offer hundreds of different models.

As a result, local suppliers largely dominate individual countries' boiler markets, and the industry's cross-border selling is much less developed than that found in most other industries. As noted by Vaillant's manager, products such as toilet cisterns and refrigerators have far less product divergence across the continent.

However, Vaillant's strategy is to focus on a few common components while producing hundreds of different types of boiler. While boilers must still be developed to meet individual countries' specifications, they will share as many common features as is logical.

Therefore, the costs of customization can be minimized without minimizing customer choice. International marketers pursuing a global strategy will need to consider how to standardize the existing product offerings and marketing activities while making unique adjustment in a local market. According to one study investigating sixteen supposedly global product attributes across three product categories in France and Malaysia, two attributes are universal and may thus be standardized. The relevance of the other fourteen attributes though is based on international market contingencies.

MARKETING OF SERVICES

Services, broadly defined, encompass all economic activities – other than agriculture, manufacturing, and mining. The service section has a great variety of industries that include: banking and insurance, travel and tourism, entertainment, wholesale and retail trade, legal and other business services, telecommunications, health care, education and training, publishing, transportation, energy, and environmental services, as well as architecture, construction, and engineering services. The impact of service can also be indirect. As countries become more developed economically, they also become more service oriented.

Firms in mature Western markets seem to gain a competitive advantage with a strong service orientation. Likewise, the evidence, while quite limited, shows that a private, more service-oriented Slovenian bank outperformed a large, older, state-supported bank.

IMPORTANCE OF SERVICES

Internationally, the "invisible" trade is responsible for one-fifth of the value of world exports. The share of commercial services in world trade has been rising. Due to information technology, communication costs will decline further. As a result, trade in services is very likely to continue to expand rapidly. In Hong Kong, the services sector has traditionally dominated the economy, accounting for 85 percent of GDP. The USA is the world's leading producer and exporter of services. The service sector is the largest component of the US economy. It accounts for 79 percent of the private sector output and 83 percent of the private non-farm employment.

The top US services exports are: tourism, transportation, financial, education, and training, business, telecommunications, equipment, entertainment, information, and health care. The major markets for American services exports, in order, are: the European Union, Japan, and Canada. Education and tourism in particular are significant earners for the USA. Some 583,000 foreign students attend US colleges and universities, and one-third or more of science and engineering degrees go to foreigners.

Likewise, travel and tourism exports are unique because buyers must travel to the USA to buy and consume the products. The top three reasons why international visitors travel to the USA are: vacation, visiting friends and relatives, and business meetings.

TYPES OF SERVICES

There are two major categories of services: consumer and business services. Business services that are exported consist of numerous and varied types, including advertising, construction and engineering, insurance, legal services, data processing, and banking.

Among the consulting and technical services are personnel training and supervision, management of facilities, and economic and business research. Services such as hairdressing are often cited as a classic example of non-tradables.

However, technological advancement has made it possible for many services to be embodied in goods that are traded internationally. Technology has also greatly benefited information-intensive services as well as knowledge-based services. The Indian software industry has been able to capture about 12 percent of the international market for customized software.

THE ECONOMIC AND LEGAL ENVIRONMENT

Like merchandise trade, exports of services are influenced by changes in relative economic conditions and exchange rates. As shown by the travel industry, when the dollar was strong in the early 1980s, the increased buying power of the dollar made foreign travel by Americans a bargain, but when the dollar weakened, foreign travelers vacationed in the USA.

The primary competition for American service firms comes from Western Europe, with new challenges being mounted by Latin American and East Asian companies. Countries such as India view service as an infant industry that must be nurtured and protected. Not surprisingly, service exports/ imports are subject to many non-tariff trade barriers. International marketers may even face outright bans on investing in certain businesses altogether. For example, foreign life insurers are not allowed to set up shop in South Korea. The National Committee for International Trade in Education publishes a report on barriers to trade in education services. The report states that American providers face numerous barriers when delivering their services abroad.

The barriers highlighted include: national legislation and policy that inhibit foreign education providers from obtaining national licenses; qualifications authorities that have difficulty recognizing foreign educational credentials; telecommunications laws that restrict the use of national satellites and receiving dishes; foreign currency controls that limit direct investment by foreign education providers, place minimum capital investment requirements on foreign-owned firms and assess prohibitively high taxes on all revenue made by foreign entities; limitations on foreign ownership; and disregard for international agreements concerning intellectual property rights.

Although service providers must abide by local laws, they may still want to try to change unfavourable laws or oppose proposed regulations that can adversely affect business activities. Since the move for fixed exchange rates will threaten the existence of the Chicago Mercantile Exchange's currency contracts, the CME established the American Coalition for Flexible Exchange Rates to present the alternative point of view to that of the advocates of fixed rates. One good sign is that services are being liberalized. One successful aspect of the Uruguay Round is the adoption of the General Agreement on Trade in Services. This agreement has extended multilateral rules to services.

MARKETING MIX AND ADAPTATION

Services have several unique characteristics; they are intangible, person-oriented, and perishable. Yet virtually all marketing concepts and strategies used to market tangible products are relevant to the marketing of services. Like a product, a company's service should also be defined broadly. American Express, for instance, does not regard itself as being in the credit-card business. Instead, the company is in the communications and information-processing business, and its computer center in Phoenix processes a quarter of a million credit-card transactions each day from all over the world. Services also require adaptation from time to time for foreign markets.

Even movies distributed abroad, more often than not, must be packaged differently. At the very least, movies distributed internationally require subtitles or overdubbing. Most Japanese were perplexed by Disney's policies of serving no alcohol and prohibiting bringing in food from outside the park. Disney, however, has made a few changes.

It added a Japanese restaurant to serve older patrons. There was no Nautilus submarine. In addition, to protect against rain and snow, more areas are covered. Service providers usually have more flexibility in providing services than products because it is more difficult for consumers to ascertain and compare the quality of services among suppliers. Their prices must still be competitive, especially when the services offered are standardized.

MARKET ENTRY STRATEGIES

Regarding market entry strategies, a service firm's unique characteristics may have some impact on entry–mode choice. In general, service firms prefer full-control modes, but firms with low asset specificity, in responding to the rising costs of integration or the diminishing ability to integrate, may have to relinquish control and seek shared-control ventures.

In practice, service firms can use virtually all market entry strategies when they are appropriate. In the financial services industry, American firms have entered into partnership and joint venture agreements with European and Japanese firms. Wells Fargo and Nikko Securities have formed a joint venture to operate a global investment management firm. Merrill Lynch and Société Générale have discussed a partnership to develop a French asset-backed securities market.

In the case of Toronto-based Four Seasons Hotels Inc., it was relatively unknown in Asia even though it manages Inn on the Park in Europe and The Pierre and The Ritz-Carlton in North America. To quickly become a dominant high-end hotelier worldwide, Four Seasons paid $122 million for a 25 percent stake in the Hong Kong-based Regent International Hotels Ltd. As part of the deal, Four Seasons also gains the rights to manage the luxury chain, thus managing forty-three hotels in seventeen countries. This acquisition strategy coupled with the management contract strategy has made it possible for Four Seasons to gain an exposure in Asia that otherwise would have taken years to develop on its own.

THE PRODUCTION OF TRADITION, MODERNITY, AND A NEW MIDDLE CLASS

Perceived as both a lure and a threat, modernity was experienced by many I grew to know in India in the early 2000s as a time, indeed, when "all that is solid melts into air." Such perceptions of modernity were intimately tied as well to visions of the past; for as scholars have long pointed out, the project of fashioning modernity is always also one of fashioning a past that is now lost or pushed away.

As we were winding our way through heavy traffic one warm Kolkata evening, I mentioned to my driver Sharfuddin Ahmad that the next day I was pleased to be visiting the daughter of a married couple living in an old age home whom I had come to know well.

I remarked that I was very interested in meeting the daughter, as I had barely ever seen any sign of the children of old-age-home residents. The daughter has a brother, I mentioned, so maybe I'll be able to find out from her why her brother doesn't look after his parents? My driver, a middle-aged man of four children approaching their marriageable years, promptly and pensively replied:

Oh, madam, these days children have become such a way that they just don't look after their parents. This age is very bad, you know? If old parents have some money, *then* the children will stay with them and look after them, as long as the parents are advantageous to them financially. But if they don't have money, after they retire, the children will just push them aside. After you give a son's marriage, you never know—What will the wife-daughterin- law be like? Will she or will she not want to look after her mother-in-law? Both those who have love marriages, and those whose parents arrange their marriages—in either case, there's no certainty what the wife will be like. And the younger generation these days prefers to live separately.

I've seen movies and things about the youth of these days—what has gotten into their heads, I just can't understand! Those who have money put their parents into some kind of an old age home, and those who don't just push their parents out somewhere into the street.

Kerala film director Jayaraj's award-winning movie, *Pathos*, portrays an aging Kerala couple, abandoned by their children who have settled in America. The couple eagerly plans for a visit from their sons, cleaning the house, preparing food, and putting up a swing in the garden for the grandchildren. Then the news comes: the children have canceled in favour of a trip to the Niagara Falls. Worse yet: the sons arrange to sell the ancestral property and place their parents in an old age home.

Director Jayaraj comments on his motivations in making the film: "In Europe, it may be normal that children leave home. But in our society, we have roots, and suddenly, all these families have started sending their children abroad; the children lose contact with their past; they forget to come home." Critiquing the contemporary materialism that has supplanted family intimacy, he adds: "These children who have settled in America are not poor. They make a lot of money and they just want to make more money".

The director reflects in another interview: "Even in a small province such as the one I live in, the number of old age homes are increasing every day. Glorified mortuaries with dead bodies of parents wait indefinitely for their children, settled abroad.

These real-life situations motivated me to make this film". On the occasion of the International Day of Older Persons on October 1, 2004, Government of India ministers published messages in newspapers around India. "We, the Indians, belonging to a glorious cultural tradition, have the inbuilt quality of respecting all the elders," begins a statement by the Minister of State for Social Justice and Empowerment. "We are the pioneers in the field of civilization and the rays of civilization have radiated throughout the world from the heart of

India," the letter goes on. Yet, the Minister cautions that "recent times" have witnessed an "erosion in the value system" that may "lead to a colossal disaster." The letter closes: "Let us make the light of love ever-shine in the evening of the lives of our elders. Let there not be any sunset in their lives." Prime Minister Manmohan Singh proclaims in his accompanying message: "We must... work to harness our traditional values of respect and reverence for the elders with existing policies."

A s is clear from these three brief vignettes, many frame their talk of contemporary aging against the backdrop of a contrast with a traditional Indian past. The ways people conceive of their past, experience the present and imagine the future are intricately wrapped together. I turn, then, to look more closely at my informants' and public media constructions of the past, as an important context for understanding the ways people are working out modes of aging in the present.

FASHIONING THE PAST

As I analyze narratives of aging in a traditional past, my aim is neither to claim nor to disclaim that this past "really was." When people speak about the past, they are always selecting certain elements from their recollections or perceptions, while neglecting to invoke others. What people construct or experience as past "tradition" *now,* only also becomes tradition in contrast to and in the present context of modernity.

So, tradition itself, like modernity, is not something that is fixed and outside history and contemporary human agency, but rather something that is being actively constructed, interpreted and used in the present.

Aging in the Joint Family: Intergenerational Reciprocity and Seva

In contemporary narratives, it is the multigenerational joint family more than anything else that represents tradition in contrast to an emerging modernity.

Although a "joint family" technically refers to a family consisting of two or more brothers living together along with their parents, spouses and children, the phrase is more loosely and popularly used in India to refer to any multigenerational household including at least one senior parent and one married adult child with spouse.

In a joint family system, old age itself is essentially a family matter, and adult children, in particular sons and daughters-in-law, live with and care for their aging parents—out of love, a deep respect for elders, and a profound sense of moral, even spiritual, duty to attempt to repay the inerasable debts they owe their parents for all the effort, expense and affection the parents expended to produce and raise them. Such joint family living pertains in contemporary narratives not only to care in old age, but also to a much wider host of social values and meanings, including supportive interdependence, fellow-feeling, a time when kinship was more important than material success, warm rambling households, connection to village lands, moral-spiritual order, patriarchy, tradition, and "Indianness."

At the heart of the joint family is a system of intergenerational reciprocity. Significantly, people indicate that it is precisely what parents once gave their children—such as a body in birth, food, material goods, money, a home, forms of love, the cleaning of urine and bowel movements—that children are expected to provide in return for their parents, years later in old age and by reconstructing and venerating them as ancestors after death. Although the common discourse is that *children* provide for and co-reside with their senior parents in these ways, in practice it is *sons* and daughters-in-law who are expected to fill this role.

Upon marriage, daughters formally relinquish obligations to their own parents, taking on responsibility for their in-laws. The providing of care for seniors in the family is often termed *seva,* service to and respect for the aged. *Seva* can be offered to deities as well as elders. When provided to elders, *seva* entails acts such as serving food and tea, massaging tired limbs, combing hair, bringing warm bath water, and offering loving respect—in short, striving to fulfill all of the elder's bodily and emotional needs.

Related to *seva* is the familiar gesture of *pranam* practiced by many Indians, when a junior bows down to touch an elder's feet in a sign of respect, and in turn the elder places his or her hands affectionately on the junior's head and offers blessings, such as "May you live well," "May your children be well." Juniors perform *seva* not simply as a gift in the present, but in exchange for the elders' earlier tremendous labors in giving birth to and fostering them. Such long-term reciprocal transactions create and sustain intimate bodily and emotional ties, what Bengalis often term *maya. Maya* means attachment, affection, compassion, and love, and Bengalis think of the ties of *maya* as entailing both bodily and emotional bindings.

A range of studies has explored how, in many contexts, Indians think of persons as relatively fluid and open, that is "dividual" or divisible in nature, in contrast to the prevalent Western notion of the relatively self-contained and bounded *in*dividual. Through transactions of food, touch, objects, words, and bodily fluids, family members give and receive parts of themselves, forging not only emotional but also bodily ties.

According to such conceptualizations, by co-residing in the same household, and by giving and receiving food, material goods, bodily care, and affection, parents and descendants not only meet each other's survival needs, but also sustain intimate ties of kinship. Property and inheritance play a somewhat complicated, ambiguous role in representations of the traditional Indian intergenerational system. Descendants are supposed to serve and provide for their elders in exchange for sacrifices their elders have already made for them. That is, elders have already earned the right to be served by their juniors, because of their tremendous earlier gifts.

However, it is widely recognized that, in practice, even in the past, the expectation of an inheritance frequently serves and served as a major motivator of filial service. Such an expectation— that those who stand to receive an inheritance from an elder are specifically obligated to provide care—becomes legally concretized in the 2007 "Maintenance and Welfare of Parents and Senior Citizens Bill".

Relatedly, one reason daughters are *not* normally expected to care for their elder parents is because daughters frequently do not stand to inherit a significant degree of their parents' property at death. Yet, even for sons, to exchange parental care for inheritance is not the "proper" way of doing things, neither in the past nor in the present. Many tales, both old and new, feature such a theme. In one classic tale, a wealthy old man preparing for his death divides his property among his sons.

The sons, having inherited all they expect to inherit, promptly begin to maltreat and neglect their father. The aged man encounters an old friend and tells him of his woes. The friend promises to help. He calls all the sons and daughters-in-law together and before them bestows to their father a large, locked, heavy chest, announcing that he is returning a long-standing debt with interest, which has now amounted to thousands of rupees. When the sons learn of their father's new wealth, they begin to be more attentive to him than ever before!

The old man carefully guards the key and is well served for the rest of his days. When he passes away, the sons and daughters-in-law greedily unlock the chest only to find it filled with stones. Importantly, the moral of such stories is never to celebrate an exchange of elder care for property, but rather to condemn the ungrateful and selfish sons and daughters-in-law who fail to recognize their normal moral obligations to care for their parents unconditionally. I n addition to intergenerational reciprocity and *seva,* "respect for elders" is also regularly presented in daily conversations and media discourse as a vital part of Indian "tradition," as conveyed in the Minister of State's message on the International Day of Older Persons cited above: "We, the Indians, belonging to a glorious cultural tradition, have the inbuilt quality of respecting all the elders." According to such discourses, all elders, not merely one's own senior kin, are to be respected, simply because of their seniority.

Thus the director of a New Delhi–based NGO, Agewell Foundation, commented to me that if given the choice between the junior Bush and the senior Bush, Indians would never have selected the junior Bush; they would have unquestionably gone with the senior. Referring to the election of George

W. Bush in 2000, he wondered why Americans had preferred the son when his father was still alive and well. He went on, speaking in English: If there is a seventy-eight-year-old person within an Indian family, most likely that person would be respected, although perhaps now relegated to the side due to modern pressures. In traditional India, the older person's opinions were sought before any family decisions were made, such as marriage or spending money. The senior-most person—of whatever age or medical condition—would be asked for his or her opinion and counsel.

This system of intergenerational reciprocity and elder respect is widely described in daily conversations and the public media as not only a culturally Indian, but also a natural, practical, and morally proper way of doing things. Narayan Sarkar, a retired engineer who at age seventy lived with his wife in their south Kolkata home, their two children settled in the United States, contemplated pensively: "In our families, we raised our children—why? Our idea, our dream was that when we grew old, our sons and daughters-in-law would serve us. And it is our dream, and a natural thing, to hope for this, to want this. We did this for our parents, and they for theirs."

A middle-aged man from rural West Bengal, who lived with and cared for his incontinent and bedridden mother, spoke similarly of the system of reciprocal parent-child care as a natural and proper one, as I came upon him one morning while he was assiduously cleaning his mother's sheets:

Caring for parents is the children's duty; it is *dharma*. As parents raised their children, children will also care for their parents during their sick years, when they get old.

For example, if I am old and I have a bowel movement, my son will clean it and he won't ask, "Why did you do it there?" This is what we did for him when he was young. When I am old and dying, who will take me to go pee and defecate? My children will have to do it.

The Indian press as well makes much of the "traditional Indian joint family" in almost every story it presents on contemporary aging, often contrasting the imagined past with an imperfect present: "India's grandparents were once revered and cared for by the extended families that they headed..."; "Famous for its culture of respect for the elderly, India is taking this tradition..."; "Earlier, the joint family system guaranteed care, concern and attention for the elderly..."; "The Great Indian Joint Family—our good old support system..."—narratives that I will complete shortly when turning to depictions of modernity. Deborah Moggach, in her 2004 novel *These Foolish Things,* floated the idea of outsourcing elderly populations from the developed world to India.

In this playful novel, British pensioners can enjoy warm weather, mango juice with their gin, inexpensive pricing, and the Indian culture of elder care in a Bangalore retirement home. "Her idea has a basis. Elderly care is part of our cultural values," an Indian government official is reported as remarking. I t is striking how contemporary representations so often highlight a very simple, idealized model of the joint family, one in which elders seamlessly receive loving respectful care, and no conflict, injustice, or economic hardship prevails. However, ethnographic research has long portrayed complex tensions in the workings of joint family relations, including conflicts between mothers-in-law and daughters-in-law, elders abandoned by children in the face of poverty, youth who ridicule the old, and disjoined nuclear-style households. Moreover, although not the most dominant narrative, some of my middle-class Kolkata and U.S-based informants did describe the Indian joint family to me in more complex and ambivalent terms, as a patriarchal, restrictive, and/or conflict-prone institution, one that could be difficult especially for daughters-in-law and even young sons.

It is notable that most of the ethnographic examples I found of statements extolling the joint family come from men, while the more ambivalent statements in my field notes were often made by women. This is not surprising, given that men have conventionally stayed within the same family from birth through death, while the position of women in both their natal and marital families is more shifting and precarious, and the difficult position of the daughter-in-law in the Indian joint family has long been a prevalent theme in public discourse. Gayatri Prasad, a middle-aged woman journalist who now lived independently in Kolkata with only her husband, her two daughters grown and married out, described the large joint family in which she had grown up.

She was the youngest of six children of her immediate parents, and they had lived with her senior uncle and his wife and children, along with her paternal grandparents. She described nostalgically how at weddings, the whole extended family would gather together for one entire month, basking in time and company, the children all having fun playing together on freshly watered lawns surrounded by flowers. In the joint family setting, she continued approvingly, she and her siblings all learned the "core Indian values"—of family closeness, spirituality, respect for elders, the merit of taking the time to help grandfather up and down the stairs without being asked. And she herself had learned to navigate the joint family, too, getting what she wanted, such as an education, by bypassing her own parents and going straight to her elder uncle. But she told also of how her sisters' lives had been "ruined" in the joint family, their personal desires and ambitions squelched. "Elders sometimes had *too* much power," Gayatri reflected.

"They were absolute dictators. They could be just and benevolent but not necessarily so." Two elder-abode residents and roommates, Kalyani-di and Uma-di, were chatting with me one evening about the changes under way in their society, notably the move from a joint family system to a system of nuclear families and old age homes. "Some people say that these changes are good," Kalyani-di reflected, "because a lot of arguing used to happen in large joint families." "When we were young girls, our mothersin- law would give us so much trouble! *Bapre bap!* " Uma-di interjected, and both women gasped and laughed for a moment as they recollected how they had suffered at the hands of their mothers-in-law.

Then Kalyani added contemplatively, with a tone of more praise than criticism, "But, people need other people to live. Before, no matter *how bad* relations were at home, people *had* to live together." Although in this and other conversations these women seemed to express a preference for what they perceived to be the joint family system of the past, they did not paint joint family living in simple harmonious terms. Whether viewed as glorious or fraught with tensions and inequalities, enduring or a waning relic of the past, it is difficult to stress enough how salient the "Indian joint family" is as a trope with which Indians interpret contemporary aging.

Throughout my fieldwork period, I found that a whole range of people—the old and young, male and female, those living in joint families and without, journalists and filmmakers—repeatedly expressed taken-for-granted assumptions that the most normal, familiar, and "traditional," even if not ubiquitous, way of aging and elder care for Indians is within a multi-generational or joint family.

I nterestingly, although so many present the joint family as a quintessentially *Indian* institution, they very often use the English "joint family" rather than an Indian-language equivalent, even when engaging in an otherwise Indian-language conversation.

In fact, other English terms frequently interjected into Bengali conversations —such as "culture" and "Indian culture"—are also, like "joint family," intended specifically to convey tradition or Indianness; but the use of English highlights how these concepts *become* specifically "traditional" and "Indian" in the context of and in contrast to the modern and foreign. So, as we will see over the following pages, "Indian culture" is frequently juxtaposed to "Western culture," and "joint family" to "nuclear family" and "old age home." Stacy Pigg notes how English terms are frequently used in Nepal to discuss taboo topics and mark them as modern. But, in a related move, Bengalis also use English terms to signal concepts that have come to be viewed as quintessentially traditional or Indian precisely when they come into contrast with perceived alternative modern or Western forms.

Aging and the Forest: Late-Life Spirituality and Human Transience

In addition to using the trope of the joint family, many Indians think of traditional ways of Indian aging in terms of the Hindu textual formulation of the four stages or *aÄramas* of life. According to the classical Hindu ethical-legal texts, persons move through a series of four life stages or "shelters"—as a student, a married householder, a disengaged forest dweller and finally a wandering renouncer. In the Hindu texts, this schema in fact applies specifically to upper-caste males, while Manu devotes little attention to defining the appropriate stages of a woman's life, which are determined by her relationships to the men upon whom she depends for support and guidance: her father in youth, her husband in marriage, and her sons in old age.

But I found that both women and men, especially among the more well-educated, frequently invoked a model of the four Hindu life stages when discussing the experiences of aging. I n the *aÄrama* schema, two life phases constitute older age.

When a man sees the sons of his sons and white hair on his head, he knows it is time to enter the forest-dweller or *vanaprastha* phase—departing from his home to live as a hermit, either with or without his wife, or remaining in the household but with a mind focused on God. The final life stage is conceptualized as a time of complete renunciation of the phenomenal world and its pleasures and ties.

As a *sannyasi* or renouncer, a man strives to become free from all worldly attachments, through taking leave of family members, abnegating caste identity, giving up all possessions, performing his own funeral rites, begging, and constantly moving from place to place so that no new attachments will develop. If a person is able to free himself from all binding attachments in this way, he may be able to attain ultimate "release" from the cycle of rebirths, redeaths, and reattachments to worldly life, or *samsar.*

Few Hindus actually move to the forest or become wandering renouncers in old age, but many do view late life as an appropriate and valuable time for focusing increasingly on God and spiritual awareness. They find compelling a model of the life course in which one concentrates on worldly matters—marriage, reproduction, and material gains and pleasures—during one's adult householder years, and then turns in later life towards spirituality and consciousness of human transience.

In some ways, though, the two models of old age explored thus far in this chapter—being served within an intimate joint family, and moving to the forest to focus on God—are in conflict, because in one a person receives care precisely within the family, and in the other the person purposefully renounces ties to the family and world. Many of the spiritually minded older Indians I know claim to find it perfectly acceptable and possible to concentrate on spirituality in late life even while living within the family.

However, as I explore in *White Saris and Sweet Mangoes,* Âothers feel conflicted and torn. They feel that their bindings of *maya*—their bodily and emotional ties to the persons, places, and things of their lived worlds—*should* decline as they grow older, but that, instead, the natural tendency is for the ties of *maya* to grow stronger and more numerous as life goes on. They *should* be loosening ties to their families and homes, developing a readiness for death and a greater awareness of God; and yet they still experience a strong yearning to see one more granddaughter's wedding, to drink one more cup of tea from a beloved grandson's hand, to eat one more sweet mango. This theme of literal or metaphorical spiritual forest-dwelling in late life is not as salient in narratives of modernity as is the theme of the joint family, possibly because the spirituality of the aged is not so widely perceived to be threatened by contemporary social changes in India.

Narratives in which the "materialist" West is encroaching into "spiritual" India—with valuable as well as injurious results—do abound; yet many feel that Western materialism exerts even more of a pull on the young than the old. Further, some of my older informants suggest that the waning family ties of modernity and globalization may in fact facilitate latelife forest-dwelling; that several features of modern social life—such as dispersed families and even old age homes—thus might actually complement certain Indian or Hindu "traditions" of aging.

IMAGINING THE PRESENT

As those I grew to know in Kolkata and the diaspora represented what they perceived to be traditional forms of Indian aging, they also produced meanings and images about a Bengali or Indian modernity. In Kolkata, modernity is perceived as entailing a cluster of concepts and terms, including the English "modern," or *adhunik* in Bengali, and other terms conveying the temporal present, such as "these days" and "nowadays" or "now". To those in Kolkata, what I gloss as "modernity" is essentially the present itself, as different from the past— "these days" as opposed to "those days," "now" as opposed to "before."

As such, modernity is regularly associated with *Western* values, lifeways and processes, and in fact, in the nineteenth century, the Bengali term most commonly employed to describe the modern or new was explicitly linked to Western education and thought, the civilization inaugurated under English rule. Modernity today is also regularly associated with features of "globalization", such as the global spread of Western values and lifeways, a global economy and media, and transnational or diasporic living.

Modernity entails as well a host of other facets of contemporary life, frequently including urban residence, nuclear families, small flats, individualism, consumerism, materialism, careerism, a persistent lack of time, weak family ties, waning patriarchy, and old age homes. Much of what Bengalis conceive as the present was inaugurated first under foreign colonial rule and then under Western-dominated globalization; this may be one reason Bengali conceptualizations of modernity cannot but be deeply ambivalent, Partha Chatterjee argues. In contrast to the "these days" felt to be not entirely of their own making, Bengalis often construct a "'those days' when there was beauty, prosperity and a healthy sociability, and which was, above all, our own creation". Amartya Sen, though, disagrees with Chatterjee's and other scholars' attempts to neatly distinguish and celebrate "our culture" as opposed to "their culture" prevalent in contemporary India. Instead, Sen argues that "In our heterogeneity and in our openness lies our pride, not our disgrace. Satyajit Ray taught us this, and that lesson is profoundly important for India. And for Asia, and for the world". Let us here look first at some of the stories of decline indeed so prevalent in India today.

As we have already begun to see, visions of modernity are often intertwined with images of aging, in which the most dominant theme is the decline of the joint family. One representative Englishlanguage newspaper story, reporting on the new Government of India legislation that could send children to jail for failing to look after their elderly parents, reads: "The Indian family scenario seems to be crumbling in the nuclear age,... with an alarming increase in the number of senior citizens taking refuge in old age homes,... city parents... hit by the empty-nest syndrome,... parallel decline in respect for the elderly". Another news story, detailing the suicide of an eighty-year-old retired Kolkata engineer whose only son lived abroad, opens: "Used to joint family structures, the elderly these days do not have even their children to talk to, leading lonely lives fraught with anguish". One Bengali gentleman reflects in a letter to the editor titled, "To Age is Now a Curse":

During our adolescence, most Bengali households consisted of joint families.... the joint families or family units with one common kitchen have disappeared like dinosaurs.... Hence now all are nuclear families.... Aged parents are neglected and ignored in the family. I am now 71 plus. Most of my relatives now live in various parts of West Bengal. I have seen the miserable condition that those of my age, and even older than me, have been thrown into!17

Acclaimed Bengali film director and actress Aparna Sen, introducing a special issue titled "Problems of Aged Parents" for the popular women's magazine *Sananda,* highlights the failed reciprocity in today's intergenerational relationships: "It takes a lot of care to bring up a human baby. What kind of a human quality is it to throw into a neglected and uncared-for state those aged parents of ours without whose care it would have been impossible for us to grow up!" The blockbuster Bollywood film, *Baghban,* one of the biggest cinematic hits in India in 2003, features the failed intergenerational reciprocity, dissolving joint family system, and chasm between the past and present envisioned as distinguishing the contemporary era. Just like a gardener who plants trees and nourishes them in the hopes that he will be able to rest in their shade when he grows old, the movie's hero, Raj Malhotra and his wife Pooja raise a family of four sons who are all now nicely settled in their professions. Against his banker's and former employer's advice, the father upon retirement invests all of his income, including his provident fund, in his sons who, he believes, are the assets who will secure his future. "I am not worried about life after retirement at all," Raj Malhotra declares towards the film's opening. "With God's grace we have four children. That means we have four invaluable fixed deposits. Eight hands are enough to support me." Once their father's bank account is empty, however, the sons and daughters-in-law treat the parents with contempt and refuse to care for them properly. In the end, the hero publishes a best-selling novel based on the mistreatment received from the hands of his sons; overnight he becomes millionaire, and the sons come groveling back, but he does not forgive them. The film closes as the father addresses a captive audience gathered to celebrate the book, his sons and daughters-in-law listening meekly:

- I am not a writer.... I have only written what life has shown me. *Baghban* is not about me or any one person. This book is about the silence that exists between the past and the future. This book is about every bridge that has broken between two successive generations. This book is about those bent shoulders on which some children sat and enjoyed the carnivals of the world. This book is about those shivering, empty hands that once held the hands of their young children and taught them how to walk.... The world has changed. Life has changed. People of my age please remember—

The father-author continues in a sarcastic tone, describing how the ways of the past have disappeared into the contemptible present:

What unworthy relationships we used to be entangled in! We saw the lord's face in our father's faces. We found heaven in the feet of our mothers. But now people have smartened up. Today's generation has become very intelligent and practical. For them every relationship is like a ladder which they step on to move up in life. And when that ladder loses its worth, then like broken furniture, broken utensils, torn clothes, yesterday's newspaper, it is discarded into a storage area.

He takes on the resonant tone of an eloquent orator:

- But life does not move up like a ladder. Life grows like a tree. Parents are *not* like the first rung in a ladder. Parents are the *roots* of the tree of life. No matter how tall the tree might grow, how healthy and green it might be, cutting the roots will not allow it to remain healthy and green.
- That is why, today, I ask with great politeness and respect, that the children for whose joy a father happily spends every penny he has earned, why do these children hesitate to shine a little light for the parents when their eyes become weak? If a father can help his son to take the first step in his life, then why can't that son help his father take the last step of his life? Why are the parents, who spend their entire lives bestowing pleasures on their children, punished with tears and loneliness?

Similar images abound in my informants' own narratives, many of whom, of diverse ages and circumstances, viewed the film *Baghban* and declared it to be highly moving, truthful, and compelling. One morning I spoke with Benu, the younger son and hard-working professional of a middle-class Kolkata family, a family still arranged jointly, housing two married brothers, their wives, children, and parents. "Today, the younger people have become very materialistic and selfish," Benu reflected.

"They only think of advancing their own careers, working hard, making money, being able to buy more things. Perhaps today they are finally able to buy a car, and then they think: Why not another? It's like that." He went on to tell of how, because of the younger generation's career orientation, shortage of time, materialism, and selfishness, "Many people don't want to care for their parents any more.

They think: I'll just put them in an old age home, or live separately. Perhaps they might even get offered an apartment as a perk through their job, and they want to accept it so they can move ahead. But perhaps their parents aren't allowed in the apartment, or there's no space for them there." He asserted that he himself, however, was totally against people "abandoning" their parents in this way and plans to live permanently as a joint family, echoing the strains of *Baghban* as he avowed: "Could a tree live without its roots? No, it would die.

My parents are my *roots*. They are the ones who raised me and nourished me and still sustain me. If they were gone, I couldn't exist—right?" I earlier introduced Kalyani-di and Uma-di, residents of a modest ten-person elder abode just a short cycle-rickshaw ride or walk from the apartment complex in which I lived with my two daughters. One evening, as the sun was setting and we were lighting mosquito coils on the small front verandah, the conversation turned, as it so often did, to how things are so different now than in the past. Kalyani-di exclaimed: "There are *huge* transformations now in our society! Everyone has become hugely selfish! No one wants to live closely any longer. Everyone wants to live alone. You'll see—as many new homes are being built, all of them are the same kind—just for two people, two people." Uma-di interjected with a mocking smile: "Soon they'll be for just one person! Even husbands and wives will live separately!" Kalyani-di had reflected on an earlier occasion:

- "If we had grown up with the idea that we might live separately from our children, then it might not be so hard to get used to now. But with our own eyes we had never seen or known anything like this. We never could have even *dreamed* that an abode for elders existed, that we would be here, in a place like this!"

Many of the older generation comment that such dramatic modern transformations are occurring with extraordinary rapidity right in their time, so that they find themselves in the unique position of having grown up in one world and now making their lives in a radically different one. "The changes have been *sudden*! Right in *our* generation!" a Dignity Foundation member exclaimed at one of their monthly gatherings. The Dignity Foundation is a "Senior Citizens Life Enrichment" association, founded in Mumbai in 1995 with branches in Kolkata, Chennai, and several other major Indian cities, its members mostly upper-middleclass seniors living independently from their children. Another gentleman jumped into the conversation, grinning, as he loudly voiced a lament: "We, of *this* generation, we suffered from our parents—how they rebuked us! And *now,* we suffer from our *children*—how *they* rebuke us!" However, as I have noted, visions of modernity or the present are richly multivalent. Not all are tales of unremitting disaster.

A more upbeat *Sunday Times of India* story claims that the joint family—essential to Indian culture—will not disappear but is rather making a resurgence, while incorporating contemporary values: "We are finally realizing the wisdom of our conventional ways. While nuclear families in the West go sub-nuclear in quest of the I-me-myself ideal, the Great Indian Joint Family is being reinvented.... Why can't we be without it?

Because unlike our western counterparts, each one of us is already more than an individual: we're a network of relationships". The newly "made-over" GIJF combines older and newer elements: members have their own space but they are there for each other, too; one can call on friends as well as family for emotional fulfillment; and one can be a working woman along with a wife, mom, daughter-in-law, and aunt. An optimistic letter to the editor from a Kolkata gentleman asserts, "A pragmatic answer to the concerns voiced would be for our society to accept and encourage the concept of old age homes as a viable replacement for the joint family of yore."

Another upbeat Sunday news story cheerfully titled "Hello! Old Age," features a group of lady friends "well into their seventies" who meet up regularly at the most modern and hip Kolkata joints, like KFC and Barista coffee. "Increasingly in Kolkata," the reporter narrates, "many senior citizens are taking charge of their own lives, wishing away dependency and redefining the conventional rules of old age."

THE SOCIAL CONTEXTS OF "MODERN AGING" IN A NEW MIDDLE CLASS

No single cause, but a whole complex of social changes, explains the emergence of novel ways of thinking about aging in India. Why *now* have old age homes suddenly burgeoned in the nation?

Why now has there been a swell of institutions geared towards producing independent, non-familybased elder living? Why now has aging taken such a hold in the public imagination as a sign of the paradoxes and complexities of the contemporary age? Several interrelated factors are at work, connected to the complex reforms of the 1990s collectively known as economic liberalization; an upwelling of money from a booming economy, outsourcing jobs, and expatriate remittances; the mounting trend of transnational migration; the substantial entry of women into the professional workforce; and the intensity of India's participation in a global circulation of ideologies.

All these processes are tied to what policy makers, academics, journalists, and the public have deemed the growth of a significant new form of middle-class life in India. The processes have clustered together to create a situation where, among the urban middle classes in particular, there is more desire to create new opportunities for managing aging other than multigenerational family residence; there is more money to pay for these new options; and yet there is anxiety about what these transformations mean—for self, family, and nation.

On the level of political economy, several forces have come together to help create a burgeoning new urban middle class with noticeably more disposable income to spend and, many believe, an array of new values and lifestyles.

The early 1990s in India marked a period of pronounced economic liberalization under the government led by Prime Minister Narasimha Rao. Since achieving status as an independent nation in 1947, India had pursued economic development with limited global entanglements; but in 1991, the economy was opened significantly to foreign investment and trade. Within five years, imports doubled, exports tripled, and foreign capital investment quintupled. Opportunities for both consumption and employment were now shaped by global markets; many professionals experienced a sharp rise in salaries; and the number of persons considered to be part of the fast-growing category of the middle class rose to somewhere between 100 and 250 million, from figures two to five times lower, over just ten years from 1989 to 1999.

In addition to bringing more disposable income into many people's hands, the new global economic climate fostered the dispersal of families across the nation and world. With the establishment of large multinational corporations in India and booming IT, biotech, and real estate industries, many young professionals were pulled from smaller or less cosmopolitan cities such as Patna and even Kolkata to careers in global economic hubs like Bangalore, New Delhi, and Mumbai—frequently leaving behind larger multigenerational homes for nuclear-family-style flats.

The past several decades have also been a period of intense transnational migration. Although people have emigrated from India for centuries, it has only been over the past few decades that it has become commonplace, even expected, for young adults of India's cosmopolitan middle classes to migrate abroad for higher education and professional opportunities, frequently ending up settling permanently overseas. In 2006, annual NRI remittances to India totaled over

$24 billion. Indians emigrate to nations such as Australia, Canada, England, Singapore, the United Arab Emirates, and very notably the United States—where the Indian American population has risen from just about 5,000 in 1960 to over 1.5 million in 2000, a three-hundred-fold increase in just forty years. In 2002 the Government of India, recognizing the cultural and economic significance of expatriate Indians, launched a Persons of Indian Origin card scheme allowing holders visa-free entry and rights to buy property and invest in the nation, aiming as well to "reinforce their emotional bonds... to their original country."

The comings and goings of NRI s play a significant role in urban middle-class Indian culture, and many of the new institutions of aging are targeted specifically at the parents of Indians living abroad. One of these is the Rosedale Gardens apartment complex constructed in 2007 on Kolkata's outskirts, with 60 of the 504 apartments designed especially for the elderly living alone, and 46 of these 60 reserved in particular for senior citizens with NRI offspring—a plan conceived by two U.S.-settled NRIs who had left parents behind in Kolkata. Media globalization has accompanied economic liberalization in India, fostering as well a powerful climate of social-cultural change and heady intermingling of local and global images and ideologies.

Transnational satellite broadcasting made its Indian debut in January 1991—cable television offerings suddenly competing with state-run television, at the same time that the slackening of foreign-exchange restrictions allowed Hollywood films to vie with Bollywood productions. Steve Derné notes the radical transformation of India's media:

- Fueled by advertisers trying to reach the new Indian market, the number of television channels grew from one state-run channel in 1991 to seventy cable channels in 1999. Access to television increased from less than 10 percent of the urban population in 1990 to nearly 75 percent by 1999. In 1991, cable television reached 300,000 homes; by 1999 it reached 24 million homes. With the easing of foreignexchange restrictions, previously unavailable Hollywood films were dubbed into Hindi and screened widely.

Elaborate shopping malls, multiplex cinemas, high-rise apartment complexes with private swimming pools, restaurants, and gyms, and international food chains like McDonalds and Pizza Hut are also mushrooming across India's metropolises, almost as one watches, where nothing like that existed even just a few years earlier.

India's first global-consumeroriented shopping mall opened in Mumbai in 1999, and within just five years according to one estimate, three hundred malls were under construction countrywide —replete with air conditioning, escalators, piped music, multiplex cinema theaters, food courts, and stores selling the leading Indian and international brands. "Earlier, a large majority of

Indians believed in the Spartan asceticism of the Father of the Nation, Mahatma Gandhi," one journalist comments on the fast-budding Indian consumer culture. "But the new generation of shoppers—like their contemporaries worldwide—believe in living for today and splurging at the mushrooming malls over the weekends". Although it is only the elite who can afford to shop regularly at chic malls, many less-well-to-do within the broad category of the middle class throng to gaze, watch a film, or indulge in snacks. A recent Kolkata *Telegraph* newspaper story reflects rather glibly on such changes in Bengali middle-class consumptive practices, linking them to even more pervasive transformations in value systems and ways of life:

- Call it a brand new attitude or a 21st century class struggle, but the *Bengali Middle Class* (BMC) is morphing. Just the other day, the quintessential 'Bong' was cocooned in his middle-class values, safely ensconced in tradition. Today, as a *Telegraph-MODE* survey reveals, the Bengali in Calcutta is all set for a dramatic metamorphosis.
- In almost every field—from clothes to food to social habits—the city's Bengali middle class is changing. Gone are the days when a BMC would subscribe to a life of restraint and modesty. Family axioms such as no late night parties, no alcohol, no impulsive decisions and no luxuries are being obliterated with remarkable ease.
- The statistics say it all. About 95 per cent of those surveyed agree that the BMC today is more modern and less conservative in outlook than it once was, while 42 per cent in the 20–50 age group care little about issues pertaining to morality. Eighty-two per cent feel it is important to be trendy and confident, and one out of every four polled loves to eat out.
- A casual visit to Pizza Hut or a *Kentucky Fried Chicken* (KFC) outlet on any given day would back the statistics. The swelling BMC crowd—which couldn't think beyond *machher jhol aar bhaat* till the other day—can be spotted tucking into crispy drums of heaven or smacking their lips after a spicy chicken and mushroom pizza.

Middle-class people in Kolkata of various ages, genders and backgrounds are talking about these matters, many finding it rather fun, new and exciting to participate in it all. At the same time, many, and not only among the older generation, engage in critical reflection, wondering if these trends are all completely desirable and if much of the Kolkata they have known and loved is fading away.

The *Telegraph* story goes on to report that a significant number of young middle-class adults has to seek mental health counseling, for as one psychiatrist explains, "When your core values receive a jolt you feel neither here nor there." And the *Telegraph* survey revealed that 91 percent in the 60–75 age group feel that the Bengali Middle Class is "going wild and losing its family values". Finally, changing gender mores form another critical dimension of contemporary middle-class culture with an impact on aging and family life. Compared to those a generation earlier, urban middle-class women are more highly educated, older at marriage, and entering the workforce in the hundreds of thousands. Such well-educated, professional women have significantly more agency in the family than had their mothers and mothers-in-law, creating an environment—as both older and younger persons, both women and men, concur—in which many women are simultaneously less inclined to and less able to co-reside with and care for their parents-in-law.

Although I encountered some persons who lamented such changes, the more pervasive sentiment at large was that the traditional position of the daughter-in-law had always left something to be desired, and that perhaps it is not so beneficial for younger women to devote their lives towards submissively serving their in-laws in the home.

It is striking that there seems today to be less public anxiety over the movement of women from the domestic to the public sphere than over the parallel movement of elders. The figures of old people and women are similar in that both can powerfully signify the family, upon which the "core values" of the nation hinge.

In the nineteenth and early twentieth centuries, during the era of British colonialism and anti-colonial nationalism, the family was held up as essential to Indianness; yet the focus then was on women. Partha Chatterjee examines how middleclass Bengali women in particular served to uphold a "traditional," Indian spiritual domain, located in the domestic sphere, distinguished from an increasingly "Western" materialist outer world that progressively engaged Indian men.

Debates linking women with family, tradition, and Indianness have also been salient within the contemporary Indian diaspora, where Indian women are widely expected to be the critical maintainers of "Indian family values," through such practices as performing culturally attuned childrearing and community service, and preparing Indian food in the home. Contemporary public anxieties, however, seem to center even more on the figure of the old person than on the woman.

This shift has transpired as middle-class women are entering the professional workforce at a dramatic rate, taking a pivotal place in the public sphere. In fact, some current discourses represent contemporary Indian women not merely as maintainers of "Indian family values," but also as significant producers of India's new global economic success.

Smitha Radhakrishnan has observed, for instance, how recent "media representations of India's booming IT industry have often carried a woman's face, suggesting that the tech revolution in India has signaled a gender revolution as well". As the roles of women are shifting, anxieties over maintaining "core

Indian values" in the home are placed increasingly onto old people and intergenerational relations. So, a cluster of events and processes have come together to foster a new kind of urban middle-class society conducive to novel ways of organizing aging and anxieties over such transitions. Some children, following career paths, have moved far from parents, across the nation or world. Some younger women are asserting preferences not to co-reside with their husband's parents.

Many parents and children feel that there is simply a different kind of "culture" at large, allowing for new lifestyles and family arrangements. There is a widespread sense, too, that a "generation gap" is increasing, partly because the pace and scope of contemporary social change seems so intense.

Although of course I argued in chapter 1 that it is misleading to think that the old cannot change, partaking in new trends such as mall culture, e-mail, and gendered egalitarianism, many of my older informants do feel that the young are moving culturally forward or away from "traditional ways" faster than they are.

That many of the younger generation engage in practices such as parties with alcohol, conspicuous consumption, fast food on the go instead of a sit-down rice meal, a busy lifestyle with little time for home, makes some elders uncomfortable.

So, if another option is now available, some elders conclude that they may actually prefer to live separately from their children—to keep relations easier—while still hopefully visiting and remaining emotionally close. One of the key reasons more people live outside of the conventional joint family today is also, simply, that they can afford to.

That is, the establishing of separate residences and the "outsourcing" of elder care from the family to, for instance, old age homes requires economic resources that earlier generations even among the fairly well-off middle classes did not readily possess.

A BRIEF INTERLUDE: THE SPHERES OF CLASS

This chapter has emphasized the perspectives and experiences of the middle classes in India, the focus of my fieldwork for this project and of recent media attention on aging.

One gets a sense from this material of a very cosmopolitan India, where people, ideas, goods, money and communications flow easily to and from a wider world.

But I also spent some time with urban poor elders in Kolkata, especially through the *Calcutta Metropolitan Institute of Gerontology* (CMIG). CMIG sought out the very poorest of the poor elderly in the city—those living in slums, on the streets, inside pipes, with or without their families— and invited them to come to the institute daily from noon to four, to sit under a ceiling fan, chat with each other, watch television, prepare paper packets for sale in the market, and receive food.

The institute was near my rented flat, and I would go when I could in the afternoons to hang out. The elderly women and men at CMIG interpreted their abject condition in old age not in terms of the changing ways of a global modernity but rather in terms of timeless poverty. Sons don't care for their parents because or when sons don't have money.

These seniors were not participating in the complex mobilities and global interconnections that so engage their middle- and upper-class neighbours. They went to and from CMIG by foot, often stooped over while carrying a small cloth bundle of meagre belongings.

- The CMIG set on which they watched their daily hour of television was tuned to a local Bengali station. They had no kin or acquaintances living abroad.
- One sultry afternoon after we had eaten a snack of *mu®i* and tea, one among the group of ladies I was sitting with asked eagerly,
- "Say, is it true, tell me, that in the U.S., people go around wearing leaves and are otherwise naked?"
- "No, no. That's not actually true," I replied.
- "OK, another thing I heard!" she continued keenly, proposing a story that seemed to her equally far-fetched. "Is it true?—that when here it's day, there it's night?"
- "Yes, that is true."
- A nother woman joined the conversation, from her spot on the floor where she rested on a piece of burlap sack. "We've seen airplanes in the sky. They're so small! Really, can people fit in them?"
- A third woman interjected to explain that when planes are on the ground, you can see that they are very large. "Have you ever been to Dum Dum?" she asked. "If you go there, you can see how big airplanes are!"
- The other ladies were interested and asked how one gets to Kolkata's airport. I explained that, in fact, a local bus goes right from the new bypass running in front of the Calcutta Metropolitan Institute of Gerontology to the airport. It would be, I thought, less than an hour's ride.
- "How much would it cost?" one of the ladies wanted to know.
- "I think about five or ten rupees," I surmised. That would be about ten or twenty cents, or the price of a half kilo of inexpensive rice.
- "See! No wonder we can't go. Where would we get money to spend like that?"
- This simple exchange brings home just how dramatically the spatial, cultural, and economic spheres of class diverge in scale.

THREAD PRODUCTION AND COSTS

MEANING OF THREAD PRODUCTION FUNCTION

In micro-economics, a thread production function is a function that specifies the output of a firm for all combinations of inputs. A meta-thread production function compares the practice of the existing entities converting inputs into output to determine the most efficient practice thread production function of the existing entities, whether the most efficient feasible practice thread production or the most efficient actual practice thread production.

In either case, the maximum output of a technologically-determined thread production process is a mathematical function of one or more inputs. Put another way, given the set of all technically feasible combinations of output and inputs, only the combinations encompassing a maximum output for a specified set of inputs would constitute the thread production function.

Alternatively, a thread production function can be defined as the specification of the minimum input requirements needed to produce designated quantities of output, given available technology. It is usually presumed that unique thread production functions can be constructed for every thread production technology.

By assuming that the maximum output technologically possible from a given set of inputs is achieved, economists using a thread production function in analysis are abstracting from the engineering and managerial problems inherently associated with a particular thread production process.

The engineering and managerial problems of technical efficiency are assumed to be solved, so that analysis can focus on the problems of allocative efficiency.

The firm is assumed to be making allocative choices concerning how much of each input factor to use and how much output to produce, given the cost of each factor, the selling price of the output, and the technological determinants represented by the thread production function.

A decision frame in which one or more inputs are held constant may be used; for example, capital may be assumed to be fixed in the short run, and labour and possibly other inputs such as raw materials variable, while in the long run, the quantities of both capital and the other factors that may be chosen by the firm are variable. In the long run, the firm may even have a choice of technologies, represented by various possible thread production functions.

The relationship of output to inputs is non-monetary; that is, a thread production function relates physical inputs to physical outputs, and prices and thread costs are reflected in the function.

But the thread production function is not a full model of the thread production process: it deliberately abstracts from inherent aspects of physical thread production processes that some would argue are essential, including error, entropy or waste. Moreover, thread production functions do not ordinarily model the business processes, either, ignoring the role of management..

The primary purpose of the thread production function is to address allocative efficiency in the use of factor inputs in thread production and the resulting distribution of income to those factors.

Under certain assumptions, the thread production function can be used to derive a marginal product for each factor, which implies an ideal division of the income generated from output into an income due to each input factor of thread production.

Thread production function as an equation

There are several ways of specifying the thread production function.

In a general mathematical form, a thread production function can be expressed as:

$$Q = f(X_1, X_2, X_3, ..., X_n)$$

where:

Q = quantity of output

$X_1, X_2, X_3, ..., X_n$ = quantities of factor inputs.

This general form does not encompass joint thread production; that is a thread production process that has multiple co-products or outputs.

One way of specifying a thread production function is simply as a table of discrete outputs and input combinations, and not as a formula or equation at all. Using an equation usually implies continual variation of output with minute variation in inputs, which is not realistic in all cases. Fixed ratios of factors, as in the case of laborers and their tools, might imply that only discrete input combinations, and therefore, discrete maximum outputs, are of practical interest.

One formulation is as a linear function:

$$Q = a + bX_1 + cX_2 + dX_3 + ...$$

where a,b,c, and d are parameters that are determined empirically.

Another is as a Cobb-Douglas thread production function:

$$Q = aX_1^b X_2^c \cdots$$

The Leontief thread production function applies to situations in which inputs must be used in fixed proportions; starting from those proportions, if usage of one input is increased without another being increased, output will not change. This thread production function is given by

$$Q = \min(aX_1, bX_2, ...).$$

Other forms include the constant elasticity of substitution thread production function, which is a generalized form of the Cobb-Douglas function, and the quadratic thread production function. The best form of the equation to use and the values of the parameters vary from company to company and industry to industry. In a short run thread production function at least one of the X's is fixed. In the long run all factor inputs are variable at the discretion of management.

Thread production function as a graph

Any of these equations can be plotted on a graph. A typical thread production function is shown in the following diagram under the assumption of a single variable input. All points above the thread production function are unobtainable with current technology, all points below are technically feasible, and all points on the function show the maximum quantity of output obtainable at the specified level of usage of the input.

From the origin, through points A, B, and C, the thread production function is rising, indicating that as additional units of inputs are used, the quantity of outputs also increases.

Beyond point C, the employment of additional units of inputs produces no additional outputs; the variable input is being used too intensively. With too much variable input use relative to the available fixed inputs, the company is experiencing negative returns to variable inputs, and diminishing total returns. In the diagram this is illustrated by the negative marginal physical product curve beyond point Z, and the declining thread production function beyond point C.

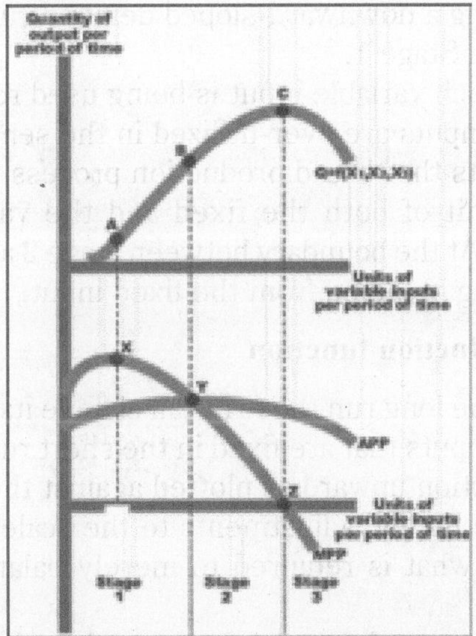

From the origin to point A, the firm is experiencing increasing returns to variable inputs. As additional inputs are employed, output increases at an increasing rate. Both marginal physical product and average physical product are rising. The inflection point A defines the point beyond which there are diminishing marginal returns, as can be seen from the declining MPP curve beyond point X. From point A to point C, the firm is experiencing positive but decreasing marginal returns to the variable input. As additional units of the input are employed, output increases but at a decreasing rate. Point B is the point beyond which there are diminishing average returns, as shown by the declining slope of the average physical product curve beyond point Y. Point B is just tangent to the steepest ray from the origin hence the average physical product is at a maximum. Beyond point B, mathematical necessity requires that the marginal curve must be below the average curve.

Stages of thread production

To simplify the interpretation of a thread production function, it is common to divide its range into 3 stages. In Stage 1 the variable input is being used with increasing output per unit, the latter reaching a maximum at point B. Because the output per unit of the variable input is improving throughout stage 1, a price-taking firm will always operate beyond this stage.

In Stage 2, output increases at a decreasing rate, and the average and marginal physical product are declining. However the average product of fixed inputs is still rising, because output is rising while fixed input usage is constant. In this stage, the employment of additional variable inputs increases the output per unit of fixed input but decreases the output per unit of the variable input. The optimum input/output combination for the price-taking firm will be in stage 2, although a firm facing a downward-sloped demand curve might find it most profitable to operate in Stage 1.

In Stage 3, too much variable input is being used relative to the available fixed inputs: variable inputs are over-utilized in the sense that their presence on the margin obstructs the thread production process rather than enhancing it. The output per unit of both the fixed and the variable input declines throughout this stage. At the boundary between stage 2 and stage 3, the highest possible output is being obtained from the fixed input.

Shifting a thread production function

By definition, in the long run the firm can change its scale of operations by adjusting the level of inputs that are fixed in the short run, thereby shifting the thread production function upward as plotted against the variable input.

If fixed inputs are lumpy, adjustments to the scale of operations may be more significant than what is required to merely balance thread production capacity with demand.

For example, you may only need to increase thread production by a million units per year to keep up with demand, but the thread production equipment upgrades that are available may involve increasing productive capacity by 2 million units per year.

If a firm is operating at a profit-maximizing level in stage one, it might, in the long run, choose to reduce its scale of operations. By reducing the amount of fixed capital inputs, the thread production function will shift down. The beginning of stage 2 shifts from B1 to B2. The profit-maximizing output level will now be in stage 2.

Homogeneous and homothetic thread production functions

There are two special classes of thread production functions that are often analyzed. The thread production function $Q = f(X_1,X_2)$ is said to be homogeneous of degree n, if given any positive constant k, $f(kX_1,kX_2) = k^n f(X_1,X_2)$. When $n > 1$, the function exhibits increasing returns to scale, and it exhibits decreasing returns to scale when $n < 1$. When it is homogeneous of degree 1, it exhibits constant returns to scale.

The presence of increasing returns means that a one percent increase in the usage levels of all inputs would result in a greater than one percent increase in output; the presence of decreasing returns means that it would result in a less than one percent increase in output. Constant returns to scale is the in-between case. In the Cobb-Douglas thread production function referred to above, returns to scale are increasing if $b + c > 1$, decreasing if $b + c < 1$, and constant if $b + c = 1$.

Homothetic functions are functions whose marginal technical rate of substitution is homogeneous of degree zero. Due to this, along rays coming from the origin, the slopes of the isoquants will be the same. Homothetic functions are of form $F(h(X_1,X_2))$ where $F(y)$ is a monotonically increasing function (the derivative of $F(y)$ is positive $(dF / dy > 0)$), and function $h(X_1,X_2)$ is a homogeneous function of any degree.

Aggregate thread production functions

In macroeconomics, aggregate thread production functions for whole nations are sometimes constructed. In theory they are the summation of all the thread production functions of individual producers; however there are methodological problems associated with aggregate thread production functions, and economists have debated extensively whether the concept is valid.

Criticisms of thread production functions

There are two major criticisms of the standard form of the thread production function. On the history of thread production functions.

On the concept of capital

During the 1950s, '60s, and '70s there was a lively debate about the theoretical soundness of thread production functions. Although the criticism was directed primarily at aggregate thread production functions, microeconomic thread production functions were also put under scrutiny.

The debate began in 1953 when Joan Robinson criticized the way the factor input capital, was measured and how the notion of factor proportions had distracted economists.

According to the argument, it is impossible to conceive of capital in such a way that its quantity is independent of the rates of interest and wages. The problem is that this independence is a precondition of constructing an isoquant. Further, the slope of the isoquant helps determine relative factor prices, but the curve cannot be constructed unless the prices are known beforehand.

Natural resources

Often natural resources are omitted from thread production functions. When Solow and Stiglitz sought to make the thread production function more realistic by adding in natural resources, they did it in a manner that economist Georgescu-Roegen criticized as a "conjuring trick" that failed to address the laws of thermodynamics. Neither Solow nor Stiglitz addressed his criticism, despite an invitation to do so in the September 1997 issue of the journal Ecological Economics.

LAW OF SUPPLY

In economics, the law of supply is the tendency of suppliers to offer more of a good at a higher price. The relationship between price and quantity supplied is usually a positive relationship. A rise in price is associated with a rise in quantity supplied. As firms produce more output, their total thread costs rise proportionately faster. The ratio of the change in total thread costs to the change in quantity is increasing. This ratio defines the firm's marginal cost of thread production. Because marginal thread costs rise and quantity produced rises, firms need to charge a higher price for each extra unit of output they produce.

ELASTICITY OF SUPPLY

Supply elasticity is defined as the percentage change in quantity supplied divided by the percentage change in price. It is calculated as per the following formula:

$$\text{Supply elasticity} = \frac{\%\Delta \text{ in quantity supplied}}{\%\Delta \text{ in price}}$$

The calculation of elasticity of supply is comparable to the calculation of elasticity of demand, except that the quantities used refer to quantities supplied instead of quantities demanded.

Factors that influence the elasticity of supply include the ability to switch to thread production of other goods, the ability to go out of business, the ability to use other resource inputs and the amount of time available to respond to a price change.

Over a short time period, firms may be able to increase output only slightly in response to an increase in prices. Over a longer period of time, the level of thread production can be adjusted greatly as thread production processes can be altered, additional workers can be hired, more plants can be built, etc. Therefore, elasticity of supply is expected to be greater with longer periods of time.

We would expect the supply elasticity of wheat to be very high as farmers can easily switch land that is used for wheat over to other crops such as corn or soybeans. On the other hand, an oil refinery cannot easily switch its thread production capacity over to another product, so low oil-refining margins do not reduce the quantity supplied by very much. Due to high capital thread costs, higher refining margins do not necessarily induce much greater supply. So the supply elasticity for oil refining is fairly low.

COST

In business, retail, and accounting, a cost is the value of money that has been used up to produce something, and hence is not available for use anymore. In economics, a cost is an alternative that is given up as a result of a decision. In business, the cost may be one of acquisition, in which case the amount of money expended to acquire it is counted as cost. In this case, money is the input that is gone in order to acquire the thing. This acquisition cost may be the sum of the cost of thread production as incurred by the original producer, and further thread costs of transaction as incurred by the acquirer over and above the price paid to the producer. Usually, the price also includes a mark-up for profit over the cost of thread production. Thread costs are often further described based on their timing or their applicability.

Accounting vs opportunity thread costs

In accounting, thread costs are the monetary value of expenditures for supplies, services, labour, products, equipment and other items purchased for use by a business or other accounting entity. It is the amount denoted on invoices as the price and recorded in bookkeeping records as an expense or asset cost basis.

Opportunity cost, also referred to as economic cost is the value of the best alternative that was not chosen in order to pursue the current endeavour—i.e, what could have been accomplished with the resources expended in the undertaking. It represents opportunities forgone. In theoretical economics, cost used without qualification often means opportunity cost.

Comparing private, external, social, and psychic thread costs

When a transaction takes place, it typically involves both private thread costs and external thread costs.

Private thread costs are the thread costs that the buyer of a good or service pays the seller. This can also be described as the thread costs internal to the firm's thread production function.

External thread costs, in contrast, are the thread costs that people other than the buyer are forced to pay as a result of the transaction. The bearers of such thread costs can be either particular individuals or society at large. Note that external thread costs are often both non-monetary and problematic to quantify for comparison with monetary values. They include things like pollution, things that society will likely have to pay for in some way or at some time in the future, but that are not included in transaction prices.

Social thread costs are the sum of private thread costs and external thread costs.

For example, the manufacturing cost of a car reflects the private cost for the manufacturer. The polluted waters or polluted air also created as part of the process of producing the car is an external cost borne by those who are affected by the pollution or who value unpolluted air or water. Because the manufacturer does not pay for this external cost and does not include this cost in the price of the car, they are said to be external to the market pricing mechanism. The air pollution from driving the car is also an externality produced by the car user in the process of using his good. The driver does not compensate for the environmental damage caused by using the car.

A psychic cost is a subset of social thread costs that specifically represent the thread costs of added stress or losses to quality of life.

Cost estimates and cost overrun

When developing a business plan for a new company, product, or project, planners typically make cost estimates in order to assess whether revenues/ benefits will cover thread costs. This is done in both business and government. Thread costs are often underestimated resulting in cost overrun during implementation. Main causes of cost underestimation and overrun are optimism bias and strategic misrepresentation. Reference class forecasting was developed to curb optimism bias and strategic misrepresentation and arrive at more accurate cost estimates. Cost Plus, is where the Price = Cost plus or minus Xper cent, where x is the percentage of built in overhead or profit margin.

COST FUNCTION

The econometrical model which is used to analyze thread costs is a model in which explanatory variable represents total thread costs and endogenous variables represent factors that influence their level. Thread production quantity is the most important factor which determines the level of total thread costs.

Total Cost

In economics, and cost accounting, total cost describes the total economic cost of thread production and is made up of variable thread costs, which vary according to the quantity of a good produced and include inputs such as labour and raw materials, plus fixed thread costs, which are independent of the quantity of a good produced and include inputs that cannot be varied in the short term, such as buildings and machinery. Total cost in economics includes the total opportunity cost of each factor of thread production in addition to fixed and variable thread costs.

The rate at which total cost changes as the amount produced changes is called marginal cost. This is also known as the marginal unit variable cost.

If one assumes that the unit variable cost is constant, as in cost-volume-profit analysis developed and used in cost accounting by the accountants, then total cost is linear in volume, and given by: total cost = fixed thread costs + unit variable cost * amount.

Fixed cost

In economics, fixed thread costs are business expenses that are not dependent on the activities of the business They tend to be time-related, such as salaries or rents being paid per month. This is in contrast to variable thread costs, which are volume-related.

In management accounting, fixed thread costs are defined as expenses that do not change in proportion to the activity of a business, within the relevant period. For example, a retailer must pay rent and utility bills irrespective of sales.

Along with variable thread costs, fixed thread costs make up one of the two components of total cost. In the most simple thread production function, total cost is equal to fixed thread costs plus variable thread costs.

Areas of confusion

Fixed thread costs should not be confused with sunk thread costs. From a pure economics perspective, fixed thread costs may not be fixed in the sense of invariate; they may change but are fixed in relation to the quantity of thread production for the relevant period. For example, a company may have unexpected and unpredictable expenses unrelated to thread production. On the other hand, thread production output may vary sharply without changing the fixed thread costs.

Strictly speaking, there is not absolute fixed cost in long-run if the relevant range is long enough. Investments in facilities, equipment, and the basic organization that can't be significantly reduced even for short periods of time without making fundamental changes are referred to as committed fixed thread costs. Discretionary fixed thread costs usually arise from annual decisions by management to spend on certain fixed cost items.

In business planning and management accounting, usage of the terms fixed thread costs, variable thread costs and others will often differ from usage in economics, and may depend on the intended use. Some cost accounting practices such as activity-based costing will allocate fixed thread costs to business activities, in effect treating them as variable thread costs. This can simplify decision-making, but can be confusing and controversial.

In accounting terminology, fixed thread costs will broadly include almost all thread costs which are not included in cost of goods sold, and variable thread costs are those captured in thread costs of goods sold. The implicit assumption required to make the equivalence between the accounting and economics terminology is that the accounting period is equal to the period in which fixed thread costs do not vary in relation to thread production. In practice, this equivalenceies does not always hold, and depending on the period under consideration by management, some overhead expenses can be adjusted by management, and the specific allocation of each expense to each category will be decided under cost accounting.

TOTAL VARIABLE COST

Total variable cost is the opportunity cost incurred in the short-run thread production that depends on the quantity of output. As the name clearly implies, total variable cost is variable, it changes. If a firm produces a little output, then total variable cost is less. If a firm produces a lot of output, then total variable cost is more.

A firm can avoid variable cost in the short run by reducing thread production to zero. This bit of information often comes in handy when the price received by a firm is so low that it falls short off covering variable cost at any positive quantity of thread production. As such, the firm might find it most "profitable" to cut losses by shutting down thread production and avoiding variable cost, until the price increases.

Variable Inputs

Total variable cost is usually, not always but usually, associated with inputs that are variable in the short run. For example, The Wacky Willy Company operates in the short run with labour as a variable input and capital as a fixed input. The cost associated with labour is a prime candidate to be a variable cost. This includes hourly wage payments to the workers and any hourly fringe benefits paid on behalf of the workers.

While labour is usually isolated as THE variable input in the short run, most short-run thread production has other variable inputs, too. The Wacky Willy Company uses an assortment of other variable inputs, all of which are part of variable cost—including cuddly cloth, squeezeably soft stuffing, and thread that make up the Stuffed Amigos; electricity needed to run the machinery; and boxes, plastic wrapping, and other packing materials used to ship the Stuffed Amigos to customers. When The Wacky Willy Company makes more Stuffed Amigos, they incur a greater cost for these variable inputs.

Cost of Stuffed Amigos

The table to the right summarizes the total variable cost of producing Wacky Willy Stuffed Amigos. The left-hand column is the quantity of Stuffed Amigos coming off the assembly line each minute, ranging from 0 to 10. For reference, the right-hand column is the total cost of producing Stuffed Amigos, the combination of total variable and total fixed cost.

The center column is the total variable cost of producing each quantity, which is exactly $0 if no Stuffed Amigos are produced and rises to $43 if 10 Stuffed Amigos are produced. This is the essence of variable cost. Greater thread production entails greater total variable cost.

- The most obvious point is that total variable cost increases with increased thread production. Producing more Stuffed Amigos means higher total variable cost. This makes sense. To produce more Stuffed Amigos, The Wacky Willy Company needs to hire more labour and buy more materials. Because these inputs incur a cost, total variable cost rises with extra thread production.
- Moreover, total variable cost is zero when no Stuffed Amigos are produced. The Wacky Willy Company incurs no variable cost when they employ no workers and buy no materials—their variable inputs.
- A last note is that the incremental increase in total variable cost is NOT the same for each quantity. In other words, total variable cost does not rise at a constant rate. This incremental change in total variable cost reflects marginal cost, a key concept in the short-run thread production and supply decision of a firm.

Total Variable Cost Curve

The total variable cost curve graphically represents the relation between total variable cost incurred by a firm in the short-run product of a good or service and the quantity produced. The total variable cost curve for Wacky Willy Stuffed Amigos thread production is illustrated in the graph to the right. Because total variable cost increases with the quantity produced, the total variable cost curve has a positive slope.

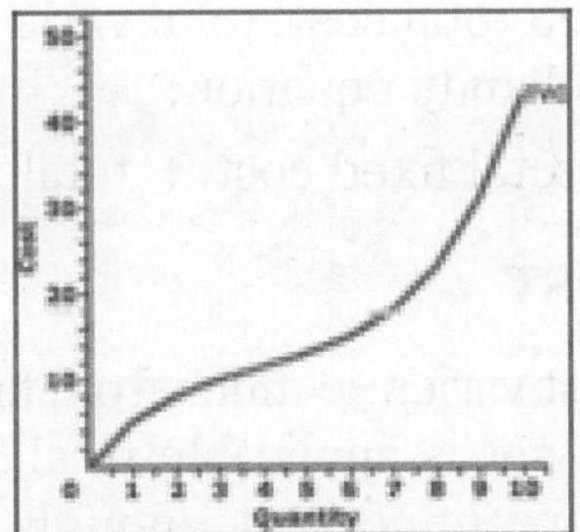

The most striking feature of the total variable cost curve is its shape. The total variable cost curve emerges from the origin, then twists and turns its way to $43. This curve begins relatively steep, then flattens, before turning increasingly steep once again.

The slope of the total variable cost curve flattens as the first four Stuffed Amigos are produced due to increasing marginal returns found in Stage I of thread production. The slope of the total variable cost curve becomes increasingly steeper after the fourth Stuffed Amigo is produced. This range of output corresponds with decreasing marginal returns, and the extremely important law of diminishing marginal returns, found in Stage II of thread production.

Average Variable Cost

Total variable cost can be used to derive a related variable cost concept—average variable cost. Average variable cost is simply variable cost per unit of output, which can be found by dividing total variable cost by the quantity of output. If, for example, total variable cost is \$43 and the quantity of output produced is 10 Stuffed Amigos, then average variable cost, that is the variable cost per unit produced, is \$4.30 (= \$43/10).

The connection between total variable cost and average variable cost is mathematically represented by this equation:

$$\text{average variable cost} = \frac{\text{total variable cost}}{\text{quantity of output}}$$

A Fixed Alternative

Total variable cost is one of two components of total cost. The other is total fixed cost. Total fixed cost is the opportunity cost incurred in the short-run thread production by a firm that does not depend on the quantity of output. A firm can produce a little output or a lot, increase or decrease thread production, or even stop producing altogether, but fixed cost remains unchanged. Total fixed cost is closely connected to the use of fixed inputs.

This means that total fixed cost is unaffected by the law of diminishing marginal returns.

The connection between total cost, total variable cost, and total fixed cost is often summarized in this handy equation:

$$\text{total cost} = \text{total fixed cost} + \text{total variable cost}$$

DETERMINANTS OF COST

The determinants of cost varies so much from firm to firm and from problem to problem that no general set is applicable to all. Nevertheless, there are a few determinants which are important enough in modern manufacturing enterprises so as to deserve special mention. They are:

- Rate of output (*i.e.*, utilization of fixed plant)
- Size of plant
- Prices of input factors (materials and labour)
- Technology
- Size of lot
- Stability of output
- Efficiency (of management as well as labour)

The word "cost" that we are using has different meanings in different settings. The kind of cost concepts to be used in a particular situation depends upon the business decision to be made. Hence an understanding of the meaning of various concepts is essential for clear business thinking.

One way of getting clear-cut distinctions among different notions of cost is to set up several alternative bases of classifying thread costs and show the relevance of each for different kinds of problems. Although there are difficulties, workable approximations can be made if these concepts of cost can be developed by having (1) a clear understanding of the management problem and of the concept of cost that is relevant for it; (2) familiarity with the business and its records (2) familiarity with the business and its records an (2=3) ingenuity and boldness. These special cost –estimates are based mainly on the accounting and statistical records of the company, though sometimes they need to be supplemented by collection of special data.

COST MANAGEMENT

Cost management is the process by which companies control and plan the thread costs of doing business. Individual projects should have customized cost management plans, and companies as a whole also integrate cost management into their overall business model. There is no single accepted definition for this term, because it has such broad applications and possible strategies. When properly implemented, cost management will translate into reduced thread costs of thread production for products and services, as well as increased value being delivered to the customer.

For a company's management to be effective overall, cost management must be an integral feature of it. It is easiest to understand this concept if it is explained in the context of a single project. For instance, before a project is started, the anticipated thread costs should be identified and measured. These expenses should then be approved before any purchasing occurs. During the process of completing a project, all incurred thread costs should be noted and kept in a record of some kind, to help ensure that the thread costs are controlled and kept in line with initial expectations, to the extent that this is possible.

Taking this approach to cost management will help a company determine whether they accurately estimated expenses at first, and will help them more closely predict expenses in the future. Any overspending can also be monitored in this way, and either eliminated in future projects or specifically approved if the expense was necessary. Cost management cannot be used in isolation; projects must be organized and tailored with this strategy in mind.

Starting a project with cost management in mind will help to avoid certain pitfalls that may be present otherwise. If the objectives of the project are not clearly defined at first, or are changed during the course of the project, cost over-runs will be more likely. If thread costs are not fully researched before the project, they may be underestimated, thereby inflating the expectation of the project's success unrealistically. Construction projects are subject to their own particular challenges; these can include constraints in the form of laws and regulations that must be planned around.

If the project is completely and clearly defined, this will facilitate effective management of the thread costs it will incur. Effective cost management strategies will help a team deliver a finished project within the allocated budget, while also making it as valuable as possible to the company. There is always the possibility of unexpected thread costs, but preparation in the form of cost management will likely make them much easier to deal with when they occur.

BREAK-EVEN ANALYSIS

Break-even analysis is a technique widely used by thread production management and management accountants. It is based on categorising thread production thread costs between those which are "variable" and those that are "fixed".

Total variable and fixed thread costs are compared with sales revenue in order to determine the level of sales volume, sales value or thread production at which the business makes neither a profit nor a loss.

The Break-Even Chart

In its simplest form, the break-even chart is a graphical representation of thread costs at various levels of activity shown on the same chart as the variation of income with the same variation in activity. The point at which neither profit nor loss is made is known as the "break-even point" and is represented on the chart below by the intersection of the two lines:

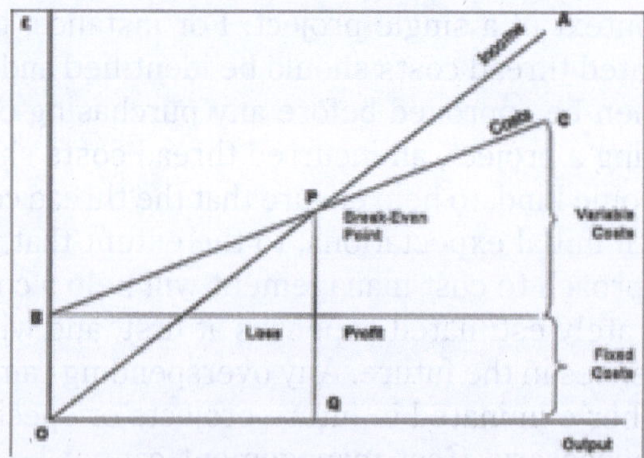

In the diagram above, the line OA represents the variation of income at varying levels of thread production activity. OB represents the total fixed thread costs in the business. As output increases, variable thread costs are incurred, meaning that total thread costs also increase. At low levels of output, Thread costs are greater than Income. At the point of intersection, P, thread costs are exactly equal to income, and hence neither profit nor loss is made.

Fixed Thread costs

Fixed thread costs are those business thread costs that are not directly related to the level of thread production or output. In other words, even if the business has a zero output or high output, the level of fixed thread costs will remain broadly the same. In the long term fixed thread costs can alter - perhaps as a result of investment in thread production capacity or through the growth in overheads required to support a larger, more complex business.

Examples of fixed thread costs:

- Rent and rates
- Depreciation
- Research and development
- Marketing thread costs
- Administration thread costs

Variable Thread costs

Variable thread costs are those thread costs which vary directly with the level of output. They represent payment output-related inputs such as raw materials, direct labour, fuel and revenue-related thread costs such as commission.

A distinction is often made between "Direct" variable thread costs and "Indirect" variable thread costs.

Direct variable thread costs are those which can be directly attributable to the thread production of a particular product or service and allocated to a particular cost centre. Raw materials and the wages those working on the thread production line are good examples.

Indirect variable thread costs cannot be directly attributable to thread production but they do vary with output. These include depreciation, maintenance and certain labour thread costs.

Semi-Variable Thread costs

Whilst the distinction between fixed and variable thread costs is a convenient way of categorising business thread costs, in reality there are some thread costs which are fixed in nature but which increase when output reaches certain levels.

These are largely related to the overall "scale" and/or complexity of the business. For example, when a business has relatively low levels of output or sales, it may not require thread costs associated with functions such as human resource management or a fully-resourced finance department. However, as the scale of the business grows then more resources are required. If thread production rises suddenly then some short-term increase in warehousing and/ or transport may be required. In these circumstances, we say that part of the cost is variable and part fixed.

PRODUCT DESCRIPTION AND APPLICATION

Sewing cotton thread is a tightly twisted thread of two or more ply that is circular when cut in cross section. It is used for industrial sewing, hand sewing and in home sewing machines. Ninety-five percent of all sewing thread that is manufactured is used in commercial and industrial sewing.

Industrial sewing thread is used as an input in shoe industry, knit wear factories, furniture and upholstery, blanket manufacturing for ribbon sewing and rural household for needle sewing and mending clothes.

Threads are wound on spools or large cones-that are marked on their ends with the size or fineness of the thread. Cotton sewing thread for hand work and machines (both home and commercial machines) has to be smooth and friction-free. Cotton thread is available in a wide range of weights, and is suitable for most sewing projects. 40wt and 50wt are the most common, but cotton threads range from 8wt to 100wt. It should be easy to thread through needles, and it should move easily when tension is applied to it. Strength to hold stitches when garments are being worn and during laundering is a requirement, as is elasticity during stitching and wear.

Cotton thread does not stretch a great deal, and will break if pulled too tightly. Cotton threads will fade with the sun, and shrink in the wash, so treat them as you would cotton fabrics. Most cotton threads sold now are mercerized. This is a chemical and heat process that increases the luster of the thread. During the mercerizing process, fuzzy threads are burned off, creating a smoother surface. This smooth surface reflects light, increasing the luster of the thread. It also has the effect of increasing water absorbency, making the thread easier to dye.

MARKET STUDY AND PLANT CAPACITY

MARKET STUDY

Present Supply and Demand

The country's requirement of cotton sewing thread is met through domestic production, and imports. Customs data reveals that there has also been export of sewing thread.

Exports, on the average accounted for about 3per cent of domestic production, while imports accounted for about 54per cent total domestic supply of sewing thread during the period 2000-2011. The total domestic supply of sewing thread, *i.e.* import and domestic production net of exports, during 2000 - 2011 is presented in Table.

Imports, domestic production, export and total apparent domestic consumption of the product averaged at 541tones, 482 tones 15 tones and 1,007,918 kg, respectively.

Apparently, imports accounted for about 54per cent total domestic supply of sewing thread.

The data depicted in Table shows a general increasing trend although it is characterized by some fluctuations. The apparent consumption of sewing thread (domestic production plus import minus export) has been increasing from period to a period. During the period 2000—2002, the yearly average apparent consumption was about 382.4 tons.

The domestic apparent consumption increased to a yearly average of 472.1 tons during the period 2003—2007. Compared to the previous three years average it is higher by about 24per cent. A huge increase on the apparent consumption is registered during the recent four years *i.e.* 2008—20011. During this period the yearly average consumption has reached to a level of 2,090 tones, which is 4.4 fold compared to the yearly average of the previous five years *i.e.* 2003—2007.

Table. Total Supply of Cotton Sweing Thread.

Year	Import[1]	Domestic Production[2]	Export[1]	Total Supply
2000	307,410	24,000	23,430	307,980
2001	518,583	7,000	-----	525,583
2002	293,738	20,000	-----	313,738
2003	803,141	1,000	30	804,111
2004	478,918	37,000	100,000	415,918
2005	377,763	108,000	11,058	474,706
2006	228,861	25,000	300	253,561
2007	363,700	49,000	560	412,140
2008	792,012	2,119,000	42	2,910,970
2009	510,441	671,000	250	1,181,191
2010	866,819	1,346,000	33,758	2,179,061
2011	948,838	1,378,866*	11,445	2,316,060

Note:

Data of domestic production of year 2011 is not published. Hence average of year 2008-2010 is assumed to be the production year of 2011.

In order to estimate the current effective domestic demand for the product a15per cent annual growth rate, which is much below the observed trend in the past, is applied by taking year 2011 as a base.

Accordingly, current domestic effective demand for sewing thread is estimated at 2,663 tones. Assuming 54per cent of the demand was satisfied by import the current unsatisfied demand for sewing thread is 1,438 tones. It should be noted that sewing thread has also an export market potential, in addition to the domestic demand.

Projected Demand

The future demand for sewing thread is a function of growth of the user industries, mainly apparel and garment manufacturing industries. Given the recent 10.6per cent rate of growth of the industrial sector (where the share of the textile industries is quite substantial) and the due attention given by the government for the textile industry, 8per cent annual rate of growth is assumed to project the demand for cotton sewing thread in Ethiopia. The total projected demand, domestic production and the unsatisfied demand is shown in Table.

Table. Demand Projection of Cotton Sewing Thread(tons).

Year	Total Projected Demand	Domestic Production	Unsatisfied Demand
2013	2,876	1,379	1,497
2014	3,106	1,379	1,727
2015	3,355	1,379	1,976
2016	3,623	1,379	2,244
2017	3,913	1,379	2,534
2018	4,226	1,379	2,847
2019	4,564	1,379	3,185
2020	4,929	1,379	3,550
2021	5,323	1,379	3,944
2022	5,749	1,379	4,370

Pricing and Distribution

Based on the recent import data obtained from Customs Authority and considering import related expenses a factory gate price of Birr 82.7 per kg is recommended.

The product will find its market outlet through the existing yarn and thread wholesale and retail channels.

PLANT CAPACITY AND PRODUCTION PROGRAMME

Plant Capacity

According to the market presented above, the unsatisfied demand for cotton sewing thread is 1497 tons in 2013 and grows to 4,370 tons by the year 2022. The envisaged plant will have a production capacity of 295 tons cotton sewing thread per year on a single shift basis.

Production can be doubled or tripled either by increasing the number of shifts or by expanding the factory.

Production Programme

The plant is expected to operate in a shift 8 hours a day for a total of 300 days a year. It is anticipated that the plant will run at 75per cent of its capacity during the first year, at 85per cent in the second year and at full capacity in the third year and then after. Production build up is made to start at reduced capacity during the initial period in order to develop substantial market outlets for the products.

Table. Production Programme.

Year	1	2	3 and above
Capacity utilization (%)	75	85	100
Total Production (ton)	221	251	295

RAW MATERIALS AND INPUTS

RAW AND AUXILIARY MATERIALS

The main raw material inputs for the production of cotton sewing threads are yarn, dyestuffs, and chemicals. Cotton yarn can be obtained locally from textile factories while dyestuff and chemicals have to be imported.

The estimated annual requirement and cost of raw material and auxiliaries inputs at a100per cent capacity utilization is given in Table.

Table. Annual Cost of Raw and Auxillary Material Inputs (tons).

Sr. No.	Description	Qty.	Cost ('000 Birr)		
			FC	LC	TC
1	Cotton yarn	337.6	-	12455.1	12455.1
2	Dye stuff	16.2	181.0	43.6	224.7
3	Chemicals	4.0	218.6	52.7	271.3
4	Cone & bobbins	1.3	221.6	-	221.6
Total		-	621.2	12551.4	13172.6

UTILITIES

The major utilities required by the plant are electricity, water and fuel. The estimated annual requirement at 100per cent capacity utilization rate and the estimated costs are given in Table.

Table. Annual Utility Requirement and Estimated Cost.

Sr. No.	Description	Qty.	Cost ('000 Birr)
1	Electric power, kWh	665,000	385.70
2	Fuel oil, liters	20,397	295.76
3	Water, m^3	10,186	101.86
Total		-	783.32

TECHNOLOGY AND ENGINEERING

TECHNOLOGY

Process Description

The basic operations involved in the manufacture of cotton sewing thread are cleaned, combed and sorted cotton is fed through a series of rollers in a process called drawing that generates a narrow band of cotton fibre. The fibre is slightly twisted to form roving and the roving is drawn and twisted again. It is spun to form a single thread that is wound and twisted with others to form the threads. Cotton sewing threads is singed over an open flame and mercerized by immersion in caustic soda. This process is strengthens the thread and give it a lustrous finish. The treated cotton sewing thread is wound or bobbins or cones.

After manufacture, the thread is dyed. Dye is mixed in large vats; several hundred colours can be produced and dye mixing is controlled by computer. Finally the dyed thread is wound on smaller spools for industrial or home use, and the spool care packed into boxes for market.

Environmental Impact Assessment

The envisage operation uses different chemicals like caustic soda and different pigments for the operation. Such operation creates an adverse effect to the environment if no proper mitigation means is considered during the design stage of the operation and hence, the right treatment means will be considered and put in place. Hence, an investment cost of Birr 500,000 is allotted for environmental impact mitigation.

ENGINEERING

Machinery and Equipment

Table. Machinery and Equipment Requirement and Estimated Cost.

Sr. No.	Description	Qty.	Cost ('000 Birr)		
			FC	LC	TC
1	Winding Machine	2	469	-	469
2	Twisting Machine	2	665	-	665
3	Reeling Machine	1	560	-	560
4	Doubling Machine	1	525	-	525
5	Dying	1	453.6	-	453.6
6	Squeezing Machine	1	203	-	203
7	Drying Machine	1	189	-	189
8	Mercerizering	1	315	-	315
9	Sizing	1	168	-	168
10	Compressor	1	490	-	490
11	Auto-process control	1	371	-	371
12	Lab-Equipment	1	210	-	210
Total			4,618.60	-	4,618.60
Insurance, Bank, Customs				1072.18	1072.18
Grand Total			4,618.60	1072.18	5,690.78

The production equipment required by the plant and their estimated costs are given in Table. All the machinery and equipment have to be imported.

Land, Building and Civil Works

The total area of land is estimated to be 2,500 m2, out of which 700 m2 will be a built-up area. The cost of building and civil works at the rate of Birr 5,000 per m2 is estimated at Birr 3,500,000.

According to the Federal Legislation on the Lease Holding of Urban Land (Proclamation No. 721/2004) in principle, urban land permit by lease is on auction or negotiation basis, however, the time and condition of applying the proclamation shall be determined by the concerned regional or city government depending on the level of development.

The legislation has also set the maximum on lease period and the payment of lease prices. The lease period ranges from 99 years for education, cultural research health, sport, NGO, religious and residential area to 80 years for industry and 70 years for trade while the lease payment period ranges from 10 years to 60 years based on the towns grade and type of investment.

Moreover, advance payment of lease based on the type of investment ranges from 5per cent to 10per cent.The lease price is payable after the grace period annually. For those that pay the entire amount of the lease will receive 0.5per cent discount from the total lease value and those that pay in installments will be charged interest based on the prevailing interest rate of banks. Moreover, based on the type of investment, two to seven years grace period shall also be provided.

However, the Federal Legislation on the Lease Holding of Urban Land apart from setting the maximum has conferred on regional and city governments the power to issue regulations on the exact terms based on the development level of each region.

In Addis Ababa, the City's Land Administration and Development Authority is directly responsible in dealing with matters concerning land. However, regarding the manufacturing sector, industrial zone preparation is one of the strategic intervention measures adopted by the City Administration for the promotion of the sector and all manufacturing projects are assumed to be located in the developed industrial zones.

Regarding land allocation of industrial zones if the land requirement of the project is below 5,000 m2, the land lease request is evaluated and decided upon by the Industrial Zone Development and Coordination Committee of the City's Investment Authority.

However, if the land request is above 5,000 m2, the request is evaluated by the City's Investment Authority and passed with recommendation to the Land Development and Administration Authority for decision, while the lease price is the same for both cases.

Moreover, the Addis Ababa City Administration has recently adopted a new land lease floor price for plots in the city. The new prices will be used as a benchmark for plots that are going to be auctioned by the city government or transferred under the new "Urban Lands Lease Holding Proclamation."

The new regulation classified the city into three zones. The first Zone is Central Market District Zone, which is classified in five levels and the floor land lease price ranges from Birr 1,686 to Birr 894 per m2. The rate for Central Market District Zone will be applicable in most areas of the city that are considered to be main business areas that entertain high level of business activities.

The second zone, Transitional Zone, will also have five levels and the floor land lease price ranges from Birr 1,035 to Birr 555 per m2.This zone includes places that are surrounding the city and are occupied by mainly residential units and industries.

The last and the third zone, Expansion Zone, is classified into four levels and covers areas that are considered to be in the outskirts of the city, where the city is expected to expand in the future. The floor land lease price in the Expansion Zone ranges from Birr 355 to Birr 191 per m2 (see Table).

Table. New Land Lease Floor Price for Plots in Addis Ababa.

Zone	Level	Floor Price/m^2
Central Market District	1st	1686
	2nd	1535
	3rd	1323
	4th	1085
	5th	894
Transitional zone	1st	1035
	2nd	935
	3rd	809
	4th	685
	5th	555
Expansion zone	1st	355
	2nd	299
	3rd	217
	4th	191

Accordingly, in order to estimate the land lease cost of the project profiles it is assumed that all new manufacturing projects will be located in industrial zones located in expansion zones. Therefore, for the profile a land lease rate of Birr 266 per m2 which is equivalent to the average floor price of plots located in expansion zone is adopted.

On the other hand, some of the investment incentives arranged by the Addis Ababa City Administration on lease payment for industrial projects are granting longer grace period and extending the lease payment period. The criterions are creation of job opportunity, foreign exchange saving, investment capital and land utilization tendency etc. Accordingly, Table shows incentives for lease payment.

Table. Incentives for Lease Payment of Industrial Projects.

Scored Point	Grace Period	Payment Completion Period	Down Payment
Above 75%	5 Years	30 Years	10%
From 50 - 75%	5 Years	28 Years	10%
From 25 - 49%	4 Years	25 Years	10%

For the purpose of this project profile, the average *i.e.* five years grace period, 28 years payment completion period and 10per cent down payment is used. The land lease period for industry is 60 years.

Accordingly, the total land lease cost at a rate of Birr 266 per m2 is estimated at Birr 665,000 of which 10per cent or Birr 66,500 will be paid in advance. The remaining Birr 598,500 will be paid in equal installments within 28 years *i.e.* Birr 21,375 annually.

NB: The land issue in the above statement narrates or shows only Addis Ababa's city administration land lease price, policy and regulations.

Accordingly the project profile prepared based on the land lease price of Addis Ababa region.

To know land lease price, police and regulation of other regional state of the country updated information is available at Ethiopian Investment Agency's web site www.eia.gov.et on the factor cost.

HUMANRESOURCE AND TRAINING REQUIREMENT

HUMANRESOURCE REQUIREMENT

The total human resource requirement of the plant is 22 persons. Details of human resource and estimated annual labour cost including fringe benefits are indicated in Table.

Table. Human Resource Requirement and Labour Cost.

Sr. No.	Description	No. Required	Salary (Birr) Monthly	Annual
1	Manager	1	5,000	60,000
2	Secretary	1	1600	19,200
3	Production Head (Supervisor)	1	3,500	42,000
4	Finance and Administration Head	1	3,500	42,000
5	Salesman	1	2,500	30,000
6	Store Keeper	1	1,600	19,200
7	Purchaser	1	2,000	24,000
8	Mechanic	1	2,400	28,800
9	Electrician	1	2,400	28,800
10	Accountant/Cashier	1	1,800	21,600
11	Driver	1	1,000	12,000
12	Production	15	24,000	288,000
13	Laborer	10	8,000	96,000
11	Guard	3	2,400	28,800
	Total	39	61,700	740,400
Employee's Benefit(20% Of Basic Salary)		-	-	148,080
Grand Total		-	-	888,480

TRAINING REQUIREMENT

The production supervisor should be given a three weeks on-the-job training by skilled technician of the equipment supplier. The cost of training is estimated at Birr 150,000.-

FINANCIAL ANALYSIS

The financial analysis of the sewing thread project is based on the data presented in the previous chapters and the following assumptions:-

- Construction period 1 year
- Source of finance 30 per cent equity and 70per cent loan
- Tax holidays 3 years
- Bank interest 10per cent
- Discount cash flow 10per cent
- Accounts receivable 30 days
- Raw material local 30 days
- Raw material imported 120 days
- Work in progress 1 day
- Finished products 30 days
- Cash in hand 5 days

- Accounts payable 30 days
- Repair and maintenance 5per cent of machinery cost

TOTAL INITIAL INVESTMENT COST

The total investment cost of the project including working capital is estimated at Birr 16.46 million (see Table). From the total investment cost, the highest share (Birr 11.21 million or 68.07per cent) is accounted by fixed investment cost followed by initial working capital (Birr 3.44 million or 20.92per cent) and pre operation cost (Birr 1.81 million or 11.00per cent). From the total investment cost, Birr 4.62 million or 28.05per cent is required in foreign currency.

Table. Intial Investment Cost.

Sr. No	Cost Items	Local Cost	Foreign Cost	Total Cost	% Share
1	Fixed investment				
1.1	Land Lease	66.50		66.50	0.40
1.2	Building and civil work	3,500.00		3,500.00	21.26
1.3	Machinery and equipment	1,072.18	4,618.60	5,690.78	34.57
1.4	Vehicles	1,500.00		1,500.00	9.11
1.5	Office furniture and equipment	450.00		450.00	2.73
	Sub total	**6,588.68**	**4,618.60**	**11,207.28**	**68.07**
2	**Pre operating cost ***				
2.1	Pre operating cost	734.54		734.54	4.46
2.2	Interest during construction	1,077.03		1,077.03	6.54
	Sub total	**1,811.57**		**1,811.57**	**11.00**
3	**Working capital ****	**3,444.39**		**3,444.39**	**20.92**
	Grand Total	**11,844.64**	**4,618.60**	**16,463.24**	**100**

Note:

*N.B Pre opertaing cost includes project implementation cost such as installation, startup, commissioning, project engineering, project management etc and capitalized intrest during construction.

** The total working capital required at full operation is Birr 4.69 million. However only the initial working capital of Birr. 3.44 million during the year of production is assumed to be fundedthrough external sources. During the remaining years the working capital requirement will be financed by funds to be generated internally.

PRODUCTION COST

The annual production cost at full operation capacity is estimated at Birr 18.94 million (see Table). The cost of raw material account for 69.57per cent of the production cost.

The other major components of the production cost are utility, depreciation, financial cost, Labour direct, marketing and distribution which account for 4.14, 9.35per cent, 3.91per cent, and 3.96per cent respectively. The remaining 9.07per cent is the share of, labour overhead and administration cost and repair and maintenance. For detail production cost see Appendix.

Table. Annual Production Cost at Full Capacity.

Items	Cost (000 Birr)	%
Raw Material and Inputs	13,173	69.57
Utilities	783	4.14
Maintenance and repair	285	1.51
Labor direct	740	3.91
Labor overheads	148	0.78
Administration Costs	250	1.32
Land lease cost	0	0.00
Cost of marketing and distribution	750	3.96
Total Operating Costs	**16,129**	**85.18**
Depreciation	1,770	9.35
Cost of Finance	1,037	5.47
Total Production Cost	**18,936**	**100.00**

FINANCIAL EVALUATION

Profitability

Based on the projected profit and loss statement, the project will generate a profit throughout its operation life. Annual net profit after tax will grow from Birr 2.77 million to Birr 4.48 million during the life of the project. Moreover, at the end of the project life the accumulated net cash flow amounts to Birr 41.00 million. For profit and loss statement and cash flow projcction see Appendix 7.A.3 and 7.A.4, respectively.

Ratios

In financial analysis, financial ratios and efficiency ratios are used as an index or yardstick for evaluating the financial position of a firm. It is also an indicator for the strength and weakness of the firm or à project. Using the year-end balance sheet figures and other relevant data, the most important ratios such as return on sales which is computed by dividing net income by revenue, return on assets (operating income divided by assets), return on equity (net profit divided by equity) and return on total investment (net profit plus interest divided by total investment) has been carried out over the period of the project life and all the results are found to be satisfactory.

Break-even Analysis

The break-even analysis establishes a relationship between operation costs and revenues. It indicates the level at which costs and revenue are in equilibrium. To this end, the break-even point for capacity utilization and sales value estimated by using income statement projection are computed as followed.

$$\text{Break Even Sales Value} = \frac{\text{Fixed Cost + Financial Cost}}{\text{Variable Margin ratio (\%)}} = \text{Birr } 9,552,060$$

$$\text{Break Even Capacity utilization} = \frac{\text{Break even Sales Value}}{\text{Sales revenue}} \times 100 = 34\,\%$$

Pay-back Period

The pay- back period, also called pay – off period is defined as the period required for recovering the original investment outlay through the accumulated net cash flows earned by the project. Accordingly, based on the projected cash flow it is estimated that the project's initial investment will be fully recovered within 3 years.

Internal Rate of Return

The internal rate of return (IRR) is the annualized effective compounded return rate that can be earned on the invested capital, *i.e.*, the yield on the investment. Put another way, the internal rate of return for an investment is the discount rate that makes the net present value of the investment's income stream total to zero. It is an indicator of the efficiency or quality of an investment. A project is a good investment proposition if its IRR is greater than the rate of return that could be earned by alternate investments or putting the money in a bank account.

Accordingly, the IRR of this project is computed to be 31.61per cent indicating the viability of the project.

Net Present Value

Net present value (NPV) is defined as the total present (discounted) value of a time series of cash flows. NPV aggregates cash flows that occur during different periods of time during the life of a project in to a common measuring unit *i.e.* present value. It is a standard method for using the time value of money to appraise long-term projects. NPV is an indicator of how much value an investment or project adds to the capital invested. In principle, a project is accepted if the NPV is non-negative.

Accordingly, the net present value of the project at 10per cent discount rate is found to be Birr 18.78 million which is acceptable. For detail discounted cash flow see Appendix 7.A.5.

ECONOMIC AND SOCIAL BENEFITS

The project can create employment for 22 persons. The project will generate Birr 11.58 million in terms of tax revenue. The establishment of such factory will have a foreign exchange saving effect to the country by substituting the current imports. The project will also create forward linkage with the shoe manufacturing, textile and furniture sub sectors and also generates other income for the Government.

Process Flow Chart and Factor Analysis in Production of a Jute Mills

Jute is the golden fibre of Bangladesh not only for the golden colour but also for the valuable contribution to the country's economy. Bangladesh holds the 2nd position as a Jute producer in the world with the average production of Jute 1.08 m ton/Year. Up to mid-twentieth century, about 80per cent of the world's jute was produced in Bangladesh and it was the country's highest foreign currency earner till early 80s. But, the emergence of petroleum-based synthetic substitutes, which were many times cheaper and convenient to use, quickly took over the market of jute. In 1980-81, jute and jute products jointly earned 68per cent of the country's total foreign exchange. The importance of jute in Bangladesh cannot be ignored. About 1.2 million farmers are still directly associated with jute cultivation. Jute sector provides about 10per cent of total employment (production, transportation, processing and marketing) in the economy. Akij jute mill is one of the biggest jute mills not only in Bangladesh but also entire world. In respect of production the mill is one of the largest in Bangladesh producing about 20,000 Metric Tons of Jute yarn annually. The mill

is creating jobs to over 4,000 people most of them are destitute females of the locality. There are six production units in this mill and capacity of each production unit is 40 metric ton yarn per day.

Process flow chart and factors that may hamper production rate are analyzed in a production unit of Akij jute mill. Flow chart is the visual representation procedure of process work and used for designing the complex processes. Production processes of jute mill from raw material to finished product are represented by the flow chart diagram. There are many different types of flowcharts, and each type has its own repertoire of boxes and notational conventions. The two most common types of boxes in a flowchart are:

- A processing step, usually called activity, and denoted as a rectangular box
- A decision usually denoted as a diamond.

In order to improve a production process, it is first necessary to understand its operation in detail. The Flowchart is a simple mapping tool that shows the sequence of actions within a process, in a form that is easy to read and communicate. The production process is a complex one that can be impacted by many factors. There are several factors inside the production unit of jute mill and there are the greater possibilities of disruption to the smooth operations or production of jute mill. Though there are both positive and negative influencing factors, it is more common to hear about and know those factors which adversely affect production. One of the primary factors which influences productivity are both intrinsic and extrinsic factors related to the employees in charge of the production process. There are many factors which can affect productivity of yarn. These can be intrinsically related to employees at an individual and interpersonal level and can be extrinsically affected by nature, acts of God and other similar uncontrollable circumstances.

Earlier some works were carried out in production of jute, defects in textile industries, global and dynamic impacts of jute, Eco- friendly jute processing etc. [1], [4]. But there is no illustration of process flow chart of jute production and factors responsible for the production. Castor oil having low content of unsaponifiable matters was chosen for the development of new jute batching emulsion processes. Different recipes for the production of emulsions were standardized in the industrial scale and their suitability was assessed according to their stability, specific gravity, temperature, viscosity and pour points Ref. [1]. Yarn manufacturing is one of the biggest manufacturing processes in textile of india (Neha Gupta, P. K. Bharti), in this paper yarn manufacturing process and defects is analyzed in textile industry Ref. [2]. Global and dynamic impact of yarn production is illustrated in (Suvalee Tangboonritruthai, Nancy L. Cassill and Willam Oxenham) paper. There several factors examination is done that affect the yarn production and consumption. This research examine trade literature and government statically reports regarding the yarn production, exports and imports, movement of machinery, preferential trade agreement and yarn consumption. This paper provides the factors that should be consider in yarn production Ref. [3]. Primary objective of this research work was to develop an alternative to mineral JBO that will be eco-friendly in addition to retaining desired properties, particularly strength and longevity. The study was conducted by processing jute with castor oil as well as the traditional mineral JBO Ref. [4].

This paper analysis the process flow chart and factors those are responsible for the decreasing of production rate in a jute mill. Production rate of jute mill can be enhanced by applying this processes, will also help to understand the common causes of production degradation.

The rest of paper is organized as follows: section 2 flow chart analyses later on factors analysis those are responsible for decreasing production rate, in section 4 linguistic terms analysis and Rest of the paper is comprised of conclusion. There is also acknowledgement and reference annexed at the last portion.

PROCESS FLOW CHART ANALYSIS

Jute is one of the most environmentally friendly 100per cent bio-degradable crops; it has many uses and thus reducing the impact on other, less sustainable natural resources. It has a cultural heritage that stretches back hundreds of years and plays a key role in the economic development of vast areas of Bangladesh as producers meet the demands of their competitive internal and export markets. Various grades of jute are used according to their quality and individual properties. Lower grade jute is often utilized as soft, protective packaging in situations where jute natural breathability is a key attribute. Jute is also used for making ropes, agricultural textiles, foods and even medicines. The most appealing application for high quality jute, however, must surely be in the production of rugs and other furnishings, where its strength and subtle iridescent shine sets it apart from the crowd.

Bangladesh produces different types of jute goods due to its worldwide demand. Process flow chart of jute production is completely different from others production like cotton production.

SELECTION OF JUTE FOR BATCH

In raw jute selection process, raw jute bales are open to figure out the defect of raw jute and remove the defective portion from the mora by the skilled workers. Raw jute is classified into two types one is 150kg weight and the other is 180 kg with or without top portion cutting. Raw jute bales are assorted according to end use like Hessian weft, sacking wrap, sacking weft etc. Raw jutes are graded according to the batch of production, jutes grade are three types mostly called A, B and C here C is the very good, B is good and A is fear. Jutes grade are used according to the quality of production regarding byers order.

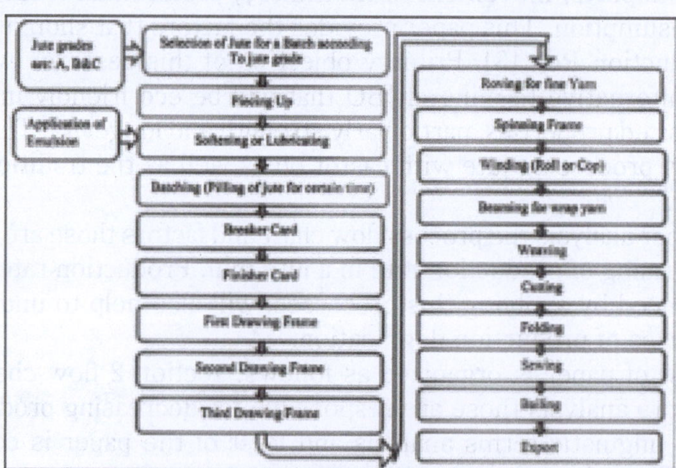

Fig. Process flow chart of jute production.

PIECING UP

At first ropes are cutting by using axe and then kept aside for treatment. Jutes are packed tightly; they are loosened by beating through a hammer. After then bulky layers are spilt into handful and each handful is 2-3lbs. various qualities of jutes and colour are mixed together. Fibres are more or less in loose so beating is not required anymore. Skilled workers separate the defective jute kept aside and those are used to produce low grade yarn. Bale cuttings are baled when the roots is wet because they are very hard and difficult to separate from the bale. After then bales are cut down or bales are broken by hammering and separated by hand.

SOFTENING OR LUBRICATING

The jutes fibres and joints in the mesh of jutes are treated through oil water emulsion in order to make the material suitable for subsequent machine. Jutes are cleaned from adhered extraneous matter during processing. The natures of emulsion are classified with type of products that are manufactured. Mineral oil is used as the softening medium for all standard jute products. The emulsion of oil and water and the pressure of fluted rollers through which it passes make fibre damp and pliable. Two methods are used for softening; use of softening machine and use of jute good spreader. Generally an emulsion plant with jute softener machine is used to lubricate and soften the bark and gummy raw jute. The emulsion plant consists of gear pump, motor, vat, jet sprayer, nozzles, emulsion tank and the jacket. In this softening process jute becomes soft and pliable and suitable for carding.

BATCHING (PILLING)

Jutes fibre is placed under a closed cover between 48-72 hours after pilling with emulsion. This produces facilitating softening of jute seeds by biological action. During piling superficial moisture penetrates inside fibre and "Thermo fillic" action take place which softener the hard portion of the root. After piling for nearly 24 hours the pile breakers carry the material to the carding machine. Generally root cutting is done after piling near the hand feed breaker carding machine. The root weight varies from 5 to 7per cent of the total weight of jute.

CARDING

Carding is a combination of operation and it is carried out to convert the long and meshy jutes into spin able fibre of desired linear density known as silvers. Carding is usually carried out in 2 or 3 stages. After 2 or 3days meshy structure of jute is passed through a sense of carding machines, which are arranged in an increasing order of fineness. There are three different carding sections:

- Breaker Carding
- Inner carding
- Finisher carding

Breaker card is the machine used to break down the meshy structure into individual long entities of filament as far as possible and also remove dust and other impurities. Breaker carding machine soften jute after piling is feed by hand in suitable weight. The machine by action with different rollers turns out raw jute in the form of jute sliver for finisher carding. In this process root cutting is necessary before feeding the material to the hand feed breaker carding machine. The product is now finer, softer and cleaner in appearance. It is then passed through the inner and finisher card is only moderately uniform, while the fibre of which it is composed of somewhat mixed up, far from being straight or parallel. Finisher carding machine make the sliver more uniform and regular in length and weight obtained from the Breaker carding machine. Finisher carding machine is identical to the Breaker carding machine, having more pair of rollers, staves, pinning arrangement and speed.

DRAWING FRAME

Drawing is a process for reducing sliver width and thickness by simultaneously mixing 4 to 6 sliver together. There are three types of Drawing Frame machine. In most mills 3 Drawing passages are used in Hessian and 2 Drawing passages are used in Sacking. The slivers obtained from finisher carding machine is fed with four slivers on to the first drawing frame machine.

- The first drawing frame machines makes blending, equalizing the sliver and doubling two or more slivers, level and provide quality and colour. This machine includes delivery roller, pressing roller, retaining roller, faller screw sliders, check spring, back spring, crimpling box etc.
- In second drawing, the Second Drawing Frame machine obtain the sliver from the First drawing machine and use six slivers and deliveries per head. The Second Drawing machine makes more uniform sliver and reduce the jute into a suitable size for third drawing.
- In the third drawing, the Third Drawing frame machine uses the sliver from second drawing. The Third Drawing machine is of high speed makes the sliver more crimpled and suitable for spinning.

Table. The Comparison of Three Drawing Process.

Drawing Process	Efficiency Range (%)	Productivity Mt/mc/shift
1st Drawing	55 - 73	1.75 – 2.2
2nd Drawing	64 - 74	1.62 – 1.9
3rd Drawing	67 - 70	1.31 – 1.4

ROVING FOR FINE YARN

The silver which is produced from the comber is thicker and not suitable to feed into the ring frame directly to produce yarn. On this case drawing frame is treated before entering the into the ring frame. There is an intermediate process is done before going to the ring frame. This process is done by roving frame. The roving frame converts the thick drawn silver into thin silver with low twist. This helps produce fine roving which is suitable to produce yarn by feeding into the ring. In this process roving frame is the input and fine roving is output. The roving is feed into the ring frame for yarn production. Roving is essential for the production of cotton yarn in case of ring spinning by ring spinning system.

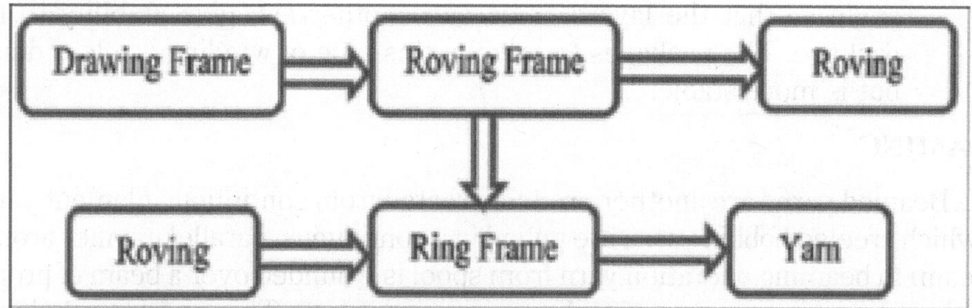

Fig. Principal for Roving Frame.

Roving frame plays an important role in the spinning process. Various types of yarn fault could be appearing for the wrong drafting or twisting. Proper roving can give better yarn properties.

SPINNING FRAME

The spinning operations carried out to produce yarn from silver, which is subjected to elongation to the specified linear density and then twisting for developing necessary yarn strength. The object of spinning and of the process that precede it is to transform the single fibres into a cohesive and workable continuous length yarn. Basically, in the case of natural fibres, the processing involves opening, blending, carding (in some cases also combing), drawing and roving to produce the material for the spinning frame. The spinning frame machine is fitted with slip draft zone and capable of producing quality yarns at high efficiency with auto-doffing arrangements also. A 4 pitch slip-draft silver frames available of 20 spindles 100 spindles, having a production range 8lbs to 28lbs with a flyer speed of 3200 to 4000 rpm. Spinning of several types of yarn is processed by spinning frame machine using different kinds of bobbins such as Food grade HFC, Sacking Wrap and Hessian Weft.

WINDING

Winding is the process of transferring yarn or thread from one type of package to another to facilitate subsequent processing. The handling of yarn is an integral part of the fibre and textile industries. Not only must the package and the yarn itself be suitable for processing on the next machine in the production process but also other factors such as packing cases, pressure due to winding tension must be considered. Basically, there are two types of winding: precision winding and drum winding.

- Precision Winding: By precision winding successive coils of yarn are laid close together in a parallel or near parallel manner. By this process it is possible to produce very dense package with maximum amount of yarn stored in a given volume.

- Non- Precision Winding: By this type of winding the package is formed by a single thread which is laid on the package at appreciable helix angle so that the layers cross one another and give stability to the package. The packages formed by this type of winding are less dense but is more stable.

BEAMING

Beamed yarns are another product created from continuous filament yarns in which creeled bobbin yarns are pulled in a continuous, parallel manner around a beam. In beaming operation yarn from spool is wounded over a beam of proper width and correct number of end to weave jute cloth. This operation helps to increase the quality of woven cloth and weaving efficiency, the wrap yarns are coated with starch paste. Adequate moisture is essential in this operation. A quality characteristic of a beam is width of beam number of ends and weight of stand and there is a continuous passage of yarn through starch solution from spools to the beam. Tamerine kernel powder (TKP) contains in water starch solution, antiseptic sodium silica fluoride (NaSiF4) and its concentration varies with the quality of yarn.

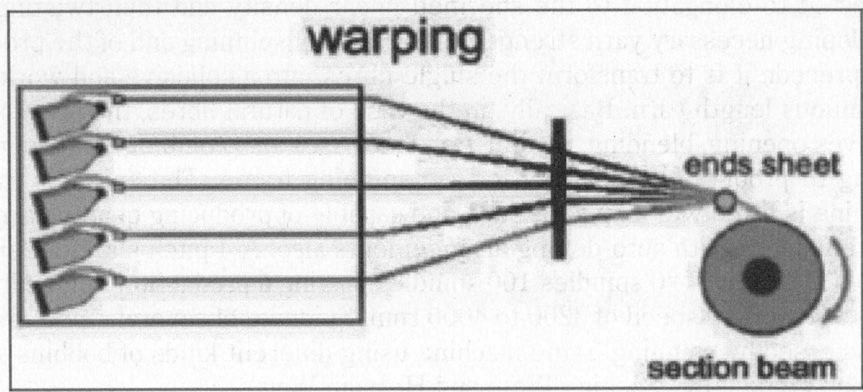

Fig. Beaming Opertion

WEAVING

Weaving is a process of interlacing two types of yarn known as warp or ends (run parallel to the weaving machine known as loom) and weft or filling yarn (run perpendicular to the loom) to produce a rigid fabric. There are separate loom for Hessian and sacking in weaving section. The Hessian looms, shuttle which contents cops can manually change. The sacking looms are equipped with eco-loader to load a cop automatically into the shuttle. In order to interlace the warp and weft yarn, there are three operations which often called primary motions are necessary.

Fig. Weaving process.

- Shedding- The process of separating the warp yarn into two layers by raising the harness to form an open area between two sets of warps and known as shed.
- Picking- The process of inserting the filling yarn through the shed by the means of the shuttle less while the shed is opening.
- Beating- The process of pushing the filling yarn into the already woven fabric at a point known as the fell and done by the reed.

CUTTING

Yarn cutting device for mechanical looms having a first blade carrier with an end-side blade and a second blade carrier with an end-side blade, at least one of said blade carriers being in an operative connection with a drive in order to carry out a yarn cutting function between the blades, wherein the drive is constructed as an electrically energized linear motor operatively connected to at least one of the blade carriers to cut yarn between the blades.

FOLDING

The folding station comprises a plurality of co-operating claws, pneumatically controlled, which fold the skein in half and push it through the center of a hollow tubular support member which has positioned around its outer surface a pre-formed paper band. Subsequent to the introduction of the folded skein into the tubular member a doffing mechanism simultaneously removes the folded skein and the pre-formed band from the tubular member, resulting in the band being positioned around the central portion of the folded skein.

The apparatus also comprises means for wrapping and sealing band forming slips around the tubular member prior to the introduction of the skein into the tubular member.

SEWING

Sewing threads are made for efficient, smooth stitching that will not break or become distorted for the life of the sewn product. Its main function is to hold together parts that could be of textile, leather etc. to form garments or other end products. Sewing machines are used to join the parts with the sewing thread in a process called Sewing. There are three types of sewing operation:

- Mechanical: stapling, sewing.
- Physical: welding or heat-setting.
- Chemical: by means of resins.

The formation of seams by physical and chemical methods is restricted to a few specialized applications, as these processes tend to alter certain properties of the textile material. Among mechanical sewing techniques, sewing maintains its prevailing position by virtue of its simplicity, sophisticated and economical production methods and the controllable elasticity of the seam produced.

BAILING

Yarn bailing process is widely appreciating clients all over the world. Such equipment is made using high grade raw material, which ensures their longer working life. Hank yarn baling press is widely used in textile industry. This press is widely appreciated for its features of superior quality, simple operation, low maintenance and durability. The resources are used to customize presses according to customer's specification.

EXPORT

Yarn is exported as per customer's requirement and quality. After achieving 100per cent customer requirement, it is allowed to export otherwise it's got back into the re-production. According to Akij jute mill rules and regulation, quality is the customer and customer is the king.

FACTOR'S ANALYSIS

Production is very important for every company to improve the global market as well as countries development. There are several factors those are directly related to the production of a jute mill as well as quality of yarn production. It is very essential to keep in mind about the control of factors in production of jute mill so that produced goods should meet customers or buyer's satisfaction. There are some qualitative factors; those are affecting the production.

RAW MATERIAL

Raw jute reeds after retting and drying are packed in the form of bales of 150 kg or 180 kg for easy transportation to jute mills. The bales from the mills godown are taken to the selection section where all the jute bales are opened to find out any defect and to remove the defective portion from the morah by experienced workers. The bales are assorted according to end use like hessian weft, sacking wrap, sacking weft, etc. After selection, jute bales are carried to softening/batching section by workers. The process of adding oil and water emulsion on jute batches is called as batching. The stack of fibre blends from different types of jute for a particular class of yarn is called a batch. The department where the jute is prepared for carding is called the batching house. In this section the fibres are conditioned by adding oil and water to it for easy processing in consequent processes. For making the jute fibre bundles suitable for next carding operation the morah prepared in selection are processed through softener or spreader machine. During its passage through these machines, oil in water emulsion is applied on jute for its moistening or lubrication.

WORKING ENVIRONMENT

When one employee is not performing at the satisfactory level, customers may notice a lack of service. This can happen for a number of reasons. It's wise to consider if the work environment at business is negatively affecting employee productivity. The ergonomics of an employee's work station also impacts performance and productivity. If desks and chairs are adjusted properly, workers are able to accomplish more. In addition, comfort levels of environmental stimuli, such as lighting and noises, can affect employee output. Another very important factor in creating a productive work environment is management's personal concern for employees. Especially relevant for employees was management's reaction to positive and negative milestones in their lives, such as births and deaths in the family.

MACHINE MAINTENANCE

Maintenance is an important factor in quality assurance, which is another basis for the successful competitive edge. Inconsistencies in equipment's lead to variability in product characteristics and result in defective parts that fail to meet the established specifications. Beyond just preventing break downs, it is necessary to keep equipment's operating within specifications (*i.e.* process capability) that will produce high level of quality.

To establish a competitive edge and to provide good customer service, companies must have reliable equipment's that will respond to customer demands when needed. Equipment's must be kept in reliable condition without costly work stoppage and down time due to repairs, if the company is to remain productive and competitive. Failure or ml-functioning of machines and equipment's in manufacturing and service industries have a direct impact on the following:

- Production capacity: Machines idled by breakdowns cannot produce, thus the capacity of the system is reduced.
- Production costs: Labour costs per unit rise because of idle labour due to machine breakdowns. When machine malfunctions result in scrap, unit labour and material costs increase
- Product and service quality: Poorly maintained equipment's produce low quality products. Equipment's that have not been properly maintained have frequent break downs and cannot provide adequate service to customers.
- Employee or customer safety: Worn-out equipment is likely to fail at any moment and these failures can cause injuries to the workers, working on those equipment's. Products such as two wheelers and

automobiles, if not serviced periodically, can break down suddenly and cause injuries to the stress.

- Customer satisfaction: When production equipment's break own, products often cannot be produced according to the master production schedules, due to work stoppages. This will lead to delayed deliveries of products to the customers.

EMPLOYER'S MOTIVATION

Unmotivated employees are likely to spend little or no effort in their jobs, avoid the workplace as much as possible, exit the organization if given the opportunity and produce low quality work. On the other hand, employees who feel motivated to work are likely to be persistent, creative and productive, turning out high quality work that they willingly undertake. Motivation is based on growth needs. It is an internal engine, and its benefits show up over a long period of time. Because the ultimate reward in motivation is personal growth. The only way to motivate an employee is to give him challenging work for which he can assume responsibility. Human motivation is so complex and so important, successful management development for the next century must include theoretical and practical education about the types of motivation, their sources, their effects on performance, and their susceptibility to various influences. Employees are the company's best assets. If employees are not as motivated, it will have a tremendous effect on productivity. The organization's overall efficiency will decline by unmotivated employees. Managers may even need to hire additional employees to complete tasks that could be done by the existing force. Proper motivation of employees is directly associated with productivity and with maintenance factors. Workers who are content with their jobs, who feel challenged, who have the opportunity to fulfill their goals will exhibit less destructive behaviour on the job. They will be absent less frequently, they will be less inclined to change jobs, and, most importantly, they will produce at a higher level.

FACILITY LAYOUT DESIGN

Facility layout and design is an important component of a business's overall operations, both in terms of maximizing the effectiveness of the production process and meeting the needs of employees. Layout planning is very important as it eliminates unnecessary costs for space and materials handling which leads to producing goods and services at a higher rate. The primary objective of selecting a layout is to minimize a function related to the travel of parts meaning the total material handling cost, the travel time and travel distance. Proper layout design helps to minimize nearly 30per cent to 40per cent of the manufacturing cost is accounted for, by materials handling. Every effort should, therefore, be made to cut down on this cost. Plant layout is a significant factor in the timely execution of orders. An ideal layout eliminates such causes of delays as shortage of space, long-distance movements of materials, spoiled work and thus contributes to the speedy execution of orders.

MATERIAL HANDLING

Material handling in addition to handling of materials in an industry is also significant in terms of costs in overall production because it is something that is quite common to all manufacturers. But when once its nature is exposed it may be difficult to overlook it as a major potential of effecting cost reduction. Material handling ranges from movement of raw material, work in progress, finished goods, rejected, scraps, packing material, etc.

These materials are of different shape and sizes as well as weight. Material handling is a systematic and scientific method of moving, packing and storing of material in appropriate and suitable location. The importance of material handling function is greater in those industries where the ratio of handling cost to the processing cost is large. Today material handling is rightly considered as one of the most potentially lucrative areas for reduction of costs in production. A properly designed and integrated material handling system provides tremendous cost saving opportunities and better production rate.

PLANNING AND SCHEDULING

A solid plan and schedule helps keep costs down and allows you to operate according to a budget with high production rate. A manufacturer must create an operations plan and schedule for the production process. Companies that have to order supplies and raw materials on a regular basis need an ordering schedule. If the company utilizes shift workers, there must also be a schedule detailing the availabilities of employees and needs of the business. High level objective of operation's planning is to decide the best way of allocation of labour and equipment as to find balance between time and use of limited resources within the organization.

An operation planning ensures that proper workflow is established by ensuring allocation of job on appropriate machines before the advent of production activities.

SAFETY MANAGEMENT

A safe work environment impacts a project's bottom line both directly and indirectly. Costs associated with incidents, including lost costs, worker's comp claims, insurance costs and legal fees are minimized in a safe work environment. A safe work environment boosts employee morale, which, in turn, increases productivity, efficiency and profit margins.

Companies can improve business operations through the use of safety procedures. Improving operations may be an unintended benefit of safety procedures. Business owners and managers that educate employees on how to best complete business functions may find new ways to improve the efficiency and effectiveness of the production process. Safety procedures may also allow employees to work quicker and improve their production output.

RESEARCH AND DEVELOPMENT (RANDD)

Research and Development (RandD) is a scientific investigation that explores the development of new goods and services, new inputs into production, new methods of producing goods and services, or new ways of operating and managing organizations. Research and Development (RandD) is a key element of many organizations and, when well-planned and used, enables a business to generate increased wealth over a period of time. The research phase includes determining product specifications, production costs and a production time line. The research also is likely to include an evaluation of the need for the product before the design begins to ensure it is a functional product that customers want to use.

PRODUCTION MANAGEMENT

Production management's responsibilities are summarized by the "five M's": men, machines, methods, materials, and money. "Men" refers to the human element in operating systems. The production manager's responsibility for materials includes the management of flow processes—both physical (raw materials) and information. The smoothness of resource movement and data flow is determined largely by the fundamental choices made in the design of the product and in the process to be used. The production manager must plan and control the process of production so that it moves smoothly at the required level of output while meeting cost and quality objectives. Process control has two purposes: first, to ensure that operations are performed according to plan, and second, to continuously monitor and evaluate the production plan to see if modifications can be devised to better meet cost, quality, delivery, flexibility, or other objectives.

LINGUISTIC VALUE

There are several factors which are directly or indirectly related with production expressed as linguistic value. Linguistic values are classified in five categories. Linguistic value is a term used in knowledge representation. It is simply knowledge in a particular field that contains no nominal value.

Table. The Linguistic Terms

Linguistic term	Value
Very Good(VG)	5
Good(G)	4
Fair(F)	3
Poor(P)	2

Relationships of factors are converted with linguistic value. Factors values are stetted in accordance with the importance of production rate. Some factors are strongly related with production and some are not, those are expressed through linguistic value.

Table. Factors in the Linguistic Value.

Factors	Linguistic Value
Raw Material(RM)	2
Working Environment(WE)	3
Machine Maintenance(MM)	3
Employer's Motivation(EM)	4
Facility Layout Design(FLD)	3
Material Handling(MH)	2
Planning & Scheduling(PS)	5
Safety Management(SM)	3
Research & Development (RD)	4
Production Management(PM)	5

This figure shows the linguistic value of factors related with production. Linguistic values are arranged in accordance with the importance of production. Factors of linguistic value expressed in a graphical method (Fig. 5); which shows the graphical representation of factors relation in production.

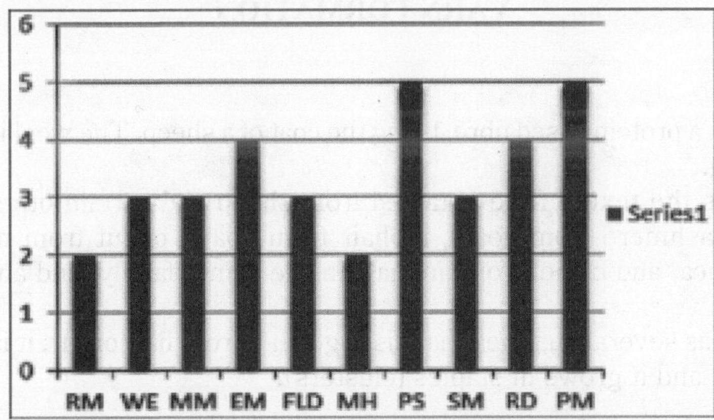

Fig. Graphical Representation of Factors.

4

Textile Manufacturing by Pre-industrial Methods

Textile manufacturing is one of the oldest human activities. The oldest known textiles date back to about 5000 B.C. In order to make textiles, the first requirement is a source of fibre from which a yarn can be made, primarily by spinning. The yarn is processed by knitting or weaving to create cloth. The machine used for weaving is the loom. Cloth is finished by what are described as wet processes to become fabric. The fabric may be dyed, printed or decorated by embroidering with coloured yarns.

The three main types of fibres are natural vegetable fibres (such as cotton, linen, jute and hemp), man-made fibres (made by industrial processes) and protein based fibres (such as wool, silk).

Almost all commercial textiles are produced by industrial methods. Textiles are still produced by pre-industrial processes in village communities in Asia, Africa and South America, as an artisan craft and a hobby in Europe and North America.

YARN FORMATION

WOOL

Wool is a protein based fibre, being the coat of a sheep. The wool is removed by shearing.

Wool is the textile fibre obtained from sheep and certain other animals, including cashmere from goats, mohair from goats, qiviut from muskoxen, vicuna, alpaca, and camel from animals in the camel family, and angora from rabbits.

Wool has several qualities that distinguish it from hair or fur: it is crimped, it is elastic, and it grows in staples (clusters).

Characteristics

Wool's scaling and crimp make it easier to spin the fleece by helping the individual fibres attach to each other, so that they stay together. Because of the crimp, wool fabrics have a greater bulk than other textiles, and retain air, which causes the product to retain heat. Insulation also works both ways; Bedouins and Tuaregs use wool clothes to keep the heat out.

The amount of crimp corresponds to the fineness of the wool fibres. A fine wool like Merino may have up to 100 crimps per inch, while the coarser wools like karakul may have as few as 1 to 2. Hair, by contrast, has little if any scale and no crimp, and little ability to bind into yarn. On sheep, the hair part of the fleece is called kemp. The relative amounts of kemp to wool vary from breed to breed, and make some fleeces more desirable for spinning, felting, or carding into batts for quilts or other insulating products.

Wool fibres are hygroscopic, meaning they readily absorb moisture. Wool fibres are hollow. Wool can absorb moisture almost one-third of its own weight. Wool absorbs sound like many other fabrics. Wool is generally a creamy white colour, although some breeds of sheep produce natural colours such as black, brown, silver, and random mixes.

Wool ignites at a higher temperature than cotton and some synthetic fibres. It has lower rate of flame spread, low heat release, low heat of combustion, and does not melt or drip; it forms a char which is insulating and self-extinguishing, and contributes less to toxic gases and smoke than other flooring products, when used in carpets. Wool carpets are specified for high safety environments, such as trains and aircraft. Wool is usually specified for garments for fire-fighters, soldiers, and others in occupations where they are exposed to the likelihood of fire.

Wool is resistant to static electricity, as the moisture retained within the fabric conducts electricity. This is why wool garments are much less likely to spark or cling to the body. The use of wool car seat covers or carpets reduces the risk of a shock when a person touches a grounded object. Wool is considered by the medical profession to be hypoallergenic.

SHEEP SHEARING

Sheep shearing, shearing or clipping is the process by which the woolen fleece of a sheep is cut off. The person who removes the sheep's wool is called a *shearer*. Typically each adult sheep is shorn once each year (a sheep may be said to have been "shorn" or "sheared", depending upon dialect). The annual shearing most often occurs in a shearing shed, a facility especially designed to process often hundreds and sometimes more than 3,000 sheep per day.

History

Shearers wear moccasins to protect their feet, grip wooden floors well, and absorb sweat.

Up until the 1870s squatters washed their sheep in nearby creeks prior to shearing. Later some expensive hot water installations were constructed on some of the larger stations for the washing. Sheep washing in Australia was influenced by the Saxony sheep breeders in Germany who washed their sheep and by the Spanish practice of washing the wool after shearing.

There were three main reasons for the custom in Australia:

1. The English manufacturers demanded that Australian woolgrowers provide their fleeces free from vegetable matter, burrs, soil, etc.
2. The dirty fleeces were hard to shear and demanded that the metal blade shears be sharpened more often.
3. Wool in Australia was carted by bullock team or horse teams and charged by weight. Washed wool was lighter and did not cost as much to transport.

The practice of washing the wool rather than the sheep evolved from the fact that hotter water could be used to wash the wool, than that used to wash the sheep. When the practice of selling wool in the grease occurred in the 1890s, wool washing became obsolete.

Australia and New Zealand had to discard the old methods of wool harvesting and evolve more efficient systems to cope with the huge numbers of sheep involved. After 1888 machine shearing was introduced, reducing second cuts and shearing time. By 1915 most large sheep station sheds in Australia had installed machines, driven by steam or later by internal combustion engines.

Shearing tables were invented in the 1950s and have not proved popular, although some are still used for crutching.

MODERN SHEARING

Today, large flocks of sheep are shorn by professional shearing teams working eight hour days, most often in spring, by machine shearing. These contractor teams will consist of shearers, shed hands and a cook (in the more isolated areas). The shed staff working hours and wages are regulated by industry awards.

A working day starts at 7:30 AM and the day is divided into four "runs" of two hours each. "Smoko" breaks of a half hour each are at 9:30 AM and again at 3:00 PM. The lunch break is taken at 12:00 PM for one hour. Most shearers are paid on a piece rate, *i.e.*, per sheep. Shearers who "tally" more than 200 sheep per day are known as "gun shearers".

Typical mass shearing of sheep today follows a well-defined workflow: remove the wool, throw the fleece onto the wool table, skirt, roll and class the fleece, place it in the appropriate wool bin, press and store the wool until it is transported.

In 1984 Australia became the last country in the world to permit the use of wide combs, due to previous Australian Workers Union rules. Although they were rare in sheds, women now take a large part in the shearing industry by working as pressers, wool rollers, rouse about, wool classers and also shearing, too.

REMOVING THE WOOL

A sheep is caught by the shearer from the catching pen and taken to his "stand" on the shearing board. It is then shorn using mechanical hand piece (see *Shearing devices* below). The wool is removed by following an efficient set of movements, devised by Godfrey Bowen in c. 1950, (the *Bowen Technique*) or the *Tally-Hi* method.

In 1963 the Tally-hi shearing system was developed by the Australian Wool Corporation and promoted using synchronized shearing demonstrations. Sheep struggle less using the Tally-Hi method, reducing strain on the shearer and there is a saving of about 30 seconds shearing each sheep. The shearer begins by removing the sheep's belly wool, which is separated from the main fleece by a rouseabout, while the sheep is still being shorn.

A professional or "gun" shearer typically removes a fleece without badly marking or cutting the sheep in two to three minutes, depending on the size and condition of the sheep, or less than two in elite competitive shearing. The shorn sheep is moved from the board via a chute in the floor, or wall, to a counting out pen, efficiently removing it from the shed.

The CSIRO in Australia has developed a non-mechanical method of shearing sheep using an injected protein that creates a natural break in the wool fibres. After fitting a retaining net to enclose the wool, sheep are injected with the protein. When the net is removed after a week, the fleece has separated and is removed by hand. In some breeds a similar process occurs naturally (see below).

SKIRTING THE FLEECE

Once the entire fleece has been removed from the sheep, the fleece is *thrown*, clean side down, on to a wool table by a shed hand (commonly known in New Zealand and Australian sheds as a *rouseabout* or *roustie*). The wool table top consists of slats spaced approximately 12 cm apart. This enables short pieces of wool, the *locks* and other debris, to gather beneath the table separately from the fleece.

The fleece is then *skirted* by one or more wool rollers to remove the sweat fribs and other less desirable parts of the fleece. The removed pieces largely consist of shorter, seeded, burry or dusty wool etc. which is still useful in the industry. As such they are placed in separate containers and sold along with fleece wool. Other items removed from the fleece on the table, such as faeces, skin fragments or twigs and leaves, are discarded a short distance from the wool table so as not to contaminate the wool and fleece.

WOOL CLASSING

Following the skirting of the fleece, it is folded, rolled and examined for its quality in a process known as wool classing, which is performed by a registered and qualified wool classer. Based on its type, the fleece is placed into the relevant wool bin ready to be pressed (mechanically compressed) when there is sufficient wool to make a wool bale.

SHEARING DEVICES

Blade Shears

Blade shears consist of two blades arranged similarly to scissors except that the hinge is at the end farthest from the point (not in the middle). The cutting edges pass each other as the shearer squeezes them together and shear the wool close to the animal's skin. Blade shears are still used today but in a more limited way. Blade shears leave some wool on a sheep and this is more suitable for cold climates where the sheep needs some protection from the elements. For those areas where no powered-machinery is available blade shears are the only option. Blades are more commonly used to shear stud rams.

Machine Shears

Machine shears, known as hand pieces, operate in a similar manner to human hair clippers in that a power-driven toothed blade, known as a cutter, is driven back and forth over the surface of a comb and the wool is cut from the animal. The original machine shears were powered by a fixed hand-crank linked to the hand piece by a shaft with only two universal joints, which afforded a very limited range of motion. Later models have more joints to allow easier positioning of the hand piece on the animal.

Electric motors on each stand have generally replaced overhead gear for driving the hand pieces. The jointed arm is replaced in many instances with a flexible shaft. Smaller motors allowed the production of shears in which the motor is in the hand piece; these are generally not used by professional shearers as the weight and heat of the motor becomes bothersome with long use.

SHEARING IN AUSTRALASIAN CULTURE

A culture has evolved out of the practice of sheep shearing, especially in post-colonial Australia and New Zealand. *Shearing the Rams*, a painting by Australian impressionist painter Tom Roberts is considered to be iconic of the livestock-growing culture or "life on the land" in Australia.

For an inversion, Michaël Leunig's *Ramming the Shears* can be seen as a sign of the shifts in Australian culture, and the extent to which the dominant rural culture is being eroded by an increasingly urban population.

The expression that Australia's wealth rode on *the sheep's back* in parts of the twentieth century no longer has the currency it once had. In 2001, Mandy Francis of Hardy's Bay, NSW constructed a black butt seat for the Street Furniture Project at Walcha, New South Wales. This seat was inspired by the combs, cutters, wool tables and grating associated with the craft and industry of shearing. During the long weekend in June 2010, 111 machine shearers and 78 blade shearers shore 6,000 Merino ewes and 178 rams at the historic 72 stand *North Tuppal* station. Along with the shearers there were 107 wool handlers and penners-up and more than 10,000 visitors to witness this event in the restored shed. Over this weekend the scene in Tom Robert's *Shearing of the Rams* was twice re-enacted for the visitors.

Many stations across Australia no longer carry sheep due to lower wool prices, drought and other disasters, but their shearing sheds remain, in a wide variety of materials and styles, and have been the subject of books and documentation for heritage authorities. Some farmers are reluctant to remove either the equipment or the sheds, and many unused sheds remain intact.

CONTESTS

Sheep shearing and wool handling competitions are held regularly in parts of the world, particularly Ireland, the UK, South Africa, New Zealand and Australia. As sheep shearing is an arduous task, speed shearers, for all types of equipment and sheep, are usually very fit and well trained. In Wales a sheep shearing contest is one of the events of the Royal Welsh Show, the country's premier agricultural show held near Builth Wells. The world's largest sheep shearing and wool handling contest, the Golden Shears, is held in the Wairarapa district, New Zealand.

SCOURING

Wool straight off a sheep, known as "greasy wool" or "wool in the grease", contains a high level of valuable lanolin, as well as dirt, dead skin, sweat residue, pesticide, and vegetable matter. Before the wool can be used for commercial purposes, it must be scoured, a process of cleaning the greasy wool. Scouring may be as simple as a bath in warm water, or as complicated as an industrial process using detergent and alkali, and specialized equipment.

In commercial wool, vegetable matter is often removed by chemical carbonization. In less processed wools, vegetable matter may be removed by hand, and some of the lanolin left intact through use of gentler detergents. This semi-grease wool can be worked into yarn and knitted into particularly water-resistant mittens or sweaters, such as those of the Aran Island fishermen. Lanolin removed from wool is widely used in cosmetic products such as hand creams.

QUALITY

The quality of wool is determined by the following factors, fibre diameter, crimp, yield, colour, and staple strength. Fibre diameter is the single most important wool characteristic determining quality and price.

Merino wool is typically 3-5 inches in length and is very fine (between 12-24 microns). The finest and most valuable wool comes from Merino hoggets.

Wool taken from sheep produced for meat is typically more coarse, and has fibres that are 1.5 to 6 inches in length. Damage or breaks in the wool can occur if the sheep is stressed while it is growing its fleece, resulting in a thin spot where the fleece is likely to break.

Wool is also separated into grades based on the measurement of the wool's diameter in microns and also its style. These grades may vary depending on the breed or purpose of the wool.

For example:

- <15.5 - Ultrafine Merino
- 15.6-18.5 - Superfine Merino
- 18.6-20 - Fine Merino
- 20.1-23 - Medium Merino
- 23< - Strong Merino
- *Comeback*: 21-26 microns, white, 90–180 mm long
- Fine crossbred: 27-31 microns, Corriedales etc.
- Medium crossbred: 32–35 microns
- *Downs*: 23-34 microns, typically lacks luster and brightness. Examples, Aussiedown, Dorset Horn, Suffolk etc.
- *Coarse crossbred*: 36> microns
- *Carpet wools*: 35-45 microns[8]

Any wool finer than 25 microns can be used for garments, while coarser grades are used for outerwear or rugs. The finer the wool, the softer it is, while coarser grades are more durable and less prone to pilling.

The finest Australian and New Zealand Merino wools are known as 1PP which is the industry benchmark of excellence for Merino wool that is 16.9 micron and finer. This style represents the top level of fineness, character, colour, and style as determined on the basis of a series of parameters in accordance with the original dictates of British Wool as applied today by the Australian Wool Exchange (AWEX) Council. Only a few dozen of the millions of bales auctioned every year can be classified and marked 1PP.[14]

HISTORY

As the raw material has been readily available since the widespread domestication of sheep - and of goats, another major provider of wool - the use of felted or woven wool for clothing and other fabrics characterizes some of the earliest civilizations. Prior to invention of shears - probably in the Iron Age - the wool was plucked out by hand or by bronze combs. The oldest known European wool textile, ca. 1500 BCE, was preserved in a Danish bog.[15] Wool fibres from wild goats found in a prehistoric cave in the Republic of Georgia as far back 34,000 BCE suggest that wool fabrics were made even earlier than this.

In Roman times, wool, linen, and leather clothed the European population; the cotton of India was a curiosity that only naturalists had heard of; and silk, imported along the Silk Road from China, was an extravagant luxury. Pliny the Elder records in his Natural History that the reputation for producing the finest wool was enjoyed by Tarentum, where selective breeding had produced sheep with a superior fleece, but which required special care.

In medieval times, as trade connections expanded, the Champagne fairs revolved around the production of wool cloth in small centers such as Proving; the network that the sequence of annual fairs developed meant that the woollens of Provins might find their way to Naples, Sicily, Cyprus, Majorca, Spain, and even Constantinople[18]. The wool trade developed into serious business, the generator of capital.

In the thirteenth century, the wool trade was the economic engine of the Low Countries and of Central Italy; by the end of the following century Italy predominated, though in the 16th century Italian production turned to silk. Both pre-industries were based on English raw wool exports - rivaled only by the sheepwalks of Castile, developed from the fifteenth century - which were a significant source of income to the English crown, which from 1275 imposed an export tax on wool called the "Great Custom".

The importance of wool to the English economy can be shown by the fact that since the 14th Century, the presiding officer of the House of Lords has sat on the "Woolsack", a chair stuffed with wool. Economies of scale were instituted in the Cistercian houses, which had accumulated great tracts of land during the twelfth and early thirteenth centuries, when land prices were low and labour still scarce.

Raw wool was baled and shipped from North Sea ports to the textile cities of Flanders, notably Ypres and Ghent, where it was dyed and worked up as cloth. At the time of the Black Death, English textile industries accounted for about 10per cent of English wool production; the English textile trade grew during the fifteenth century, to the point where export of wool was discouraged.

Over the centuries, various British laws controlled the wool trade or required the use of wool even in burials. The smuggling of wool out of the country, known as owling, was at one time punishable by the cutting off of a hand. After the Restoration, fine English woollens began to compete with silks in the international market, partly aided by the Navigation Acts; in 1699 English crown forbade its American colonies to trade wool with anyone but England herself.

A great deal of the value of woolen textiles was in the dyeing and finishing of the woven product. In each of the centers of the textile trade, the manufacturing process came to be subdivided into a collection of trades, overseen by an entrepreneur in a system called by the English the "putting-out" system, or "cottage industry", and the Verlagssystem by the Germans.

In this system of producing wool cloth, until recently perpetuated in the production of Harris tweeds, the entrepreneur provides the raw materials and an advance, the remainder being paid upon delivery of the product. Written contracts bound the artisans to specified terms. Fernand Braudel traces the appearance of the system in the thirteenth-century economic boom, quoting a document of 1275[18] The system effectively by-passed the guilds' restrictions.

Before the flowering of the Renaissance, the Medici and other great banking houses of Florence had built their wealth and banking system on their textile industry based on wool, overseen by the Arte della Lana, the wool guild: wool textile interests guided Florentine policies. Francesco Datini, the "merchant of Prato", established in 1383 an Arte della Lana for that small Tuscan city.

The sheepwalks of Castile shaped the landscape and the fortunes of the meseta that lies in the heart of the Iberian peninsula; in the sixteenth century, a unified Spain allowed export of Merino lambs only with royal permission. The German wool market - based on sheep of Spanish origin - did not overtake British wool until comparatively late.

Australia's colonial economy was based on sheep raising, and the Australian wool trade eventually overtook that of the Germans by 1845, furnishing wool for Bradford, which developed as the heart of industrialized woollens production. A World War I era poster sponsored by the United States Department of Agriculture encouraging children to raise sheep to provide needed war supplies.

Due to decreasing demand with increased use of synthetic fibres, wool production is much less than what it was in the past. The collapse in the price of wool began in late 1966 with a 40per cent drop; with occasional interruptions, the price has tended down.

The result has been sharply reduced production and movement of resources into production of other commodities, in the case of sheep growers, to production of meat.

Superwash wool (or washable wool) technology first appeared in the early 1970s to produce wool that has been specially treated so that it is machine washable and may be tumble-dried. This wool is produced using an acid bath that removes the "scales" from the fibre, or by coating the fibre with a polymer that prevents the scales from attaching to each other and causing shrinkage. This process results in a fibre that holds longevity and durability over synthetic materials, while retaining its shape.

In December 2004, a bale of the world's finest wool, averaging 11.8 micron, sold for $3,000 per kilogram at auction in Melbourne, Victoria. This fleece wool tested with an average yield of 74.5per cent, 68 mm long, and had 40 newtons per kilotex strength.

The result was $AUD279,000 for the bale.[24] The finest bale of wool ever auctioned sold for a seasonal record of 269,000 cents per kilo during June 2008. This bale was produced by the Hillcreston Pinehill Partnership and measured 11.6 microns, 72.1per cent yield and had a 43 Newtons per kilotex strength measurement. The bale realised $247,480 and was exported to India.

During 2007 a new wool suit was developed and sold in Japan that can be washed in the shower, and dries off ready to wear within hours with no ironing required. The suit was developed using Australian Merino wool and it enables woven products made from wool, such as suits, trousers and skirts, to be cleaned using a domestic shower at home. In December 2006 the General Assembly of the United Nations proclaimed 2009 to be the International Year of Natural Fibres, so to raise the profile of wool and other natural fibres.

PRODUCTION

Global wool production is approximately 1.3 million tonnes per year, of which 60per cent goes into apparel. Australia is the leading producer of wool which is mostly from Merino sheep.

New Zealand is the second-largest producer of wool, and the largest producer of crossbred wool. China is the third-largest producer of wool. Breeds such as Lincoln, Romney, Tukidale, Drysdale and Elliotdale produce coarser fibres, and wool from these sheep is usually used for making carpets.

In the United States, Texas, New Mexico and Colourado have large commercial sheep flocks and their mainstay is the Rambouillet (or French Merino). There is also a thriving home-flock contingent of small-scale farmers who raise small hobby flocks of specialty sheep for the hand spinning market. These small-scale farmers offer a wide selection of fleece.

Global woolclip (total amount of wool shorn) 2004/2005:

- Australia: 25per cent of global woolclip (475 million kg greasy, 2004/2005)
- China: 18per cent
- New Zealand: 11per cent
- Argentina: 3per cent
- Turkey: 2per cent
- Iran: 2per cent
- United Kingdom: 2per cent
- India: 2per cent
- Sudan: 2per cent
- South Africa: 1per cent
- United States: 0.77per cent

Organic wool is becoming more and more popular. This wool is very limited in supply and much of it comes from New Zealand and Australia. It is becoming easier to find in clothing and other products, but these products often carry a higher price.

Wool is environmentally preferable (as compared to petroleum-based Nylon or Polypropylene) as a material for carpets as well, in particular when combined with a natural binding and the use of formaldehyde-free glues. Animal rights groups have noted issues with the production of wool, such as Mulesing.

MARKETING

About 85per cent of wool sold in Australia is sold by open cry auction. Sale by Sample is a method in which a mechanical claw takes a sample from each bale in a line or lot of wool. These grab samples are bulked, objectively measured, and a sample of not less than 4 kg is displayed in a box for the buyer to examine. The Australian Wool Exchange (AWEX) conducts sales primarily in Sydney, Melbourne, Newcastle, and Fremantle. There are about 80 brokers and agents throughout Australia.

About 7per cent of Australian wool is sold by private treaty on farms or to local wool-handling facilities. This option gives wool growers benefit from reduced transport, warehousing, and selling costs. This method is preferred for small lots or mixed butts in order to make savings on reclassing and testing. About 5per cent of Australian wool is sold over the internet on an electronic offer board.

This option gives wool growers the ability to set firm price targets, reoffer passed in wool and offer lots to the market quickly and efficiently. This method works well for tested lots as buyers use these results to make a purchase. 97per cent of wool is sold without sample inspection however as of dec 2009, 59per cent of wool listed had been passed in from auction. Growers through certain brokers can allocate their wool to a sale and what price their wool will be reserved at. Sale by tender can achieve considerable cost savings on wool clips large enough to make it worthwhile for potential buyers to submit tenders. Some marketing firms sell wool on a consignment basis, obtaining a fixed percentage as commission.

Forward selling: Some buyers offer a secure price for forward delivery of wool based on estimated measurements or the results of previous clips. Prices are quoted at current market rates and are locked in for the season. Premiums and discounts are added to cover variations in micron, yield, tensile strength, etc., which are confirmed by actual test results when available.Another method of selling wool includes sales direct to wool mills.

The British Wool Marketing Board operates a central marketing system for UK fleece wool with the aim of achieving the best possible net returns for farmers. Less than half of New Zealand's wool is sold at auction, while around 45per cent for farmers sell wool directly to private buyers and end-users. Some businesses in New Zealand like Blue House Yarns have turned to selling organic wool, a new trend on wool production.

United States sheep producers' market wool with private or cooperative wool warehouses, but wool pools are common in many states. In some cases, wool is pooled in a local market area but sold through a wool warehouse. Wool offered with objective measurement test results is preferred. Imported apparel wool and carpet wool goes directly to central markets where it is handled by the large merchants and manufacturers.

USES

In addition to clothing, wool has been used for blankets, horse rugs, saddle cloths, carpeting, felt, wool insulation (also see links) and upholstery. Wool felt covers piano hammers, and it is used to absorb odours and noise in heavy machinery and stereo speakers. Ancient Greeks lined their helmets with felt, and Roman legionnaires used breastplates made of wool felt.

Wool has also been traditionally used to cover cloth diapers. Wool fibre exteriors are hydrophobic (repel water) and the interior of the wool fibre is hygroscopic (attracts water); this makes a wool garment able to cover a wet diaper while inhibiting wicking, so outer garments remain dry. Wool felted and treated with lanolin is water resistant, air permeable, and slightly antibacterial, so it resists the buildup of odour. Some modern cloth diapers use felted wool fabric for covers, and there are several modern commercial knitting patterns for wool diaper covers.

Initial studies of woollen underwear have found it prevented heat and sweat rashes because it more readily absorbs the moisture than other fibres.Merino wool has been used in baby sleep products such as swaddle baby wrap blankets and infant sleeping bags. Wool is an animal protein, and as such it can be used as a soil fertilizer, being a slow release source of nitrogen and ready made amino acids.

VIRGIN WOOL

Wool spun for the first time is called Virgin wool.

SHODDY OR RECYCLED WOOL:

It is made by cutting or tearing apart existing wool fabric and respinning the resulting fibres. As this process makes the wool fibres shorter, the remanufactured fabric is inferior to the original. The recycled wool may be mixed with raw wool, wool noil, or another fibre such as cotton to increase the average fibre length. Such yarns are typically used as weft yarns with a cotton warp. This process was invented in the Heavy Woollen District of West Yorkshire and created a micro-economy in this area for many years.

Ragg is a sturdy wool fibre made into yarn and used in many rugged applications like gloves. Worsted is a strong, long-staple, combed wool yarn with a hard surface.

Woollen is a soft, short-staple, carded wool yarn typically used for knitting. In traditional weaving, woollen weft yarn (for softness and warmth) is frequently combined with a worsted warp yarn for strength on the loom.

EVENTS

A buyer of Merino wool, Ermenegildo Zegna, has offered awards for Australian wool producers. In 1963, the first Ermenegildo Zegna Perpetual

Trophy was presented in Tasmania for growers of "Superfine skirted Merino fleece". In 1980, a national award, the Ermenegildo Zegna Trophy for Extrafine Wool Production, was launched. In 2004, this award became known as the Ermenegildo Zegna Unprotected Wool Trophy. In 1998, an Ermenegildo Zegna Protected Wool Trophy was launched for fleece from sheep coated for around nine months of the year.

In 2002, the Ermenegildo Zegna Vellus Aureum Trophy was launched for wool that is 13.9 micron and finer. Wool from Australia, New Zealand, Argentina, and South Africa may enter, and a winner is named from each country. In April 2008, New Zealand won the Ermenegildo Zegna Vellus Aureum Trophy for the first time with a fleece that measured 10.8 microns. This contest awards the winning fleece weight with the same weight in gold as a prize, hence the name.

In 2010 an ultra-fine, 10 micron fleece, from Windradeen, near Pyramul, New South Wales set a new world record in the fineness of wool fleeces when it won the Ermenegildo Zegna Vellus Aureum International Trophy.

Since 2000, Loro Piana has awarded a cup for the world's finest bale of wool that produces just enough fabric for 50 tailor-made suits. The prize is awarded to an Australian or New Zealand wool grower who produces the year's finest bale.

The New England Merino Field days which display local studs, wool, and sheep are held during January, every two years (in even numbered years) around the Walcha, New South Wales district. The Annual Wool Fashion Awards, which showcase the use of Merino wool by fashion designers, are hosted by the city of Armidale, New South Wales in March each year.

This event encourages young and established fashion designers to display their talents. During each May, Armidale hosts the annual New England Wool Expo to display wool fashions, handicrafts, demonstrations, shearing competitions, yard dog trials, and more.

In July, the annual Australian Sheep and Wool Show is held in Bendigo, Victoria.

This is the largest sheep and wool show in the world, with goats and alpacas as well as woolcraft competitions and displays, fleece competitions, sheepdog trials, shearing, and wool handling.

The largest competition in the world for objectively-measured fleeces is the Australian Fleece Competition, which is held annually at Bendigo. In 2008, there were 475 entries from all states of Australia with first and second prizes going to the Northern Tablelands, New South Wales fleeces.

WOOL CLASSING

Wool classing is an occupation for which people are trained to produce uniform, predictable, low risk lines of wool. This is carried out by examining the characteristics of the wool in its raw state.

The characteristics which a wool classer would examine are:

- *Breed of the sheep*: Shedding breeds will increase the risk of medulated and/or pigmented fibres. Any sheep likely to have dark fibres should be shorn last to avoid contamination. The age of the sheep will have a bearing on the fibre diameter and value of wool, too.
- *Chemical usage*: Ensure that all rules have been followed.
- *Brands, seedy jowls* and *shanks*: Must be removed from fleeces and broken.
- *Stain*: Must be removed from bellies and fleeces and identified in a separate line.
- *Wool crimp*: The number of bends per unit length along the wool fibre approximately indicates spinning capacity of the wool. fibres with a fine crimp have many bends and usually have a small diameter. Such fibre can be spun into fine yarns, with great lengths of yarn for a given weight of wool, and greater market value. Fine fibres may be utilised in the production of fine garments such as men's suits whereas the coarser fibres may be used for the production of carpet and other sturdy products. Crimp is measured in crimps per inch or crimps per centimetre. Average diameter or mean fibre diameter is measured in micrometres (microns). For generations, English wool-handlers categorized wool along the above lines estimating spinning capacity by eye and touch. This spread worldwide as the Bradford system.
- *Wool Strength* (also known as *tensile strength*) determines wool's ability to withstand processing. Weaker wools produce more waste in carding and spinning. Weaker wools may be used for production of felt, or combined with other fibres, etc.
- *Wool colour*: Indicates whether wool is able to be dyed in light shades. Colour may be graded depending upon the natural colour, impurities and various stains present. Severely stained wool decreases prices dramatically. However, it is difficult to assess colour accurately without proper measurement, since some stains will wash out in the processing, whereas others are quite persistent.

The fleece is skirted to remove excess frib, seed and burr etc to leave the fleece as reasonably even as possible in good respects. The parts of wool taken from a sheep are graded separately. The fleece forming the bulk of the yield is placed with other fleece wool as the main line, other pieces such as the neck, belly and skirtings (inferior wool from edges) are sold for such purposes where the shorter wools are required (for example: fillings, carpets, insulation).

Whilst in some places crimp may determine which grade the fleece will be placed into, this subjective assessment is not always reliable and processors prefer that wools are measured objectively by qualified laboratories. Some of the superfine wool growers do in shed wool testing, but this can only be used as a guide. This enables wool classers to place wool into lines of a consistent quality.

A shedhand, known as a wool presser, places the wool into approved wool packs in a wool press to produce a bale of wool that must meet regulations concerning its fastenings, length, weight and branding if it is to be sold at auction in Australasia. All Merino fleece wool sold at auction in Australia is objectively measured for fibre diameter, yield (including the amount of vegetable matter), staple length, staple strength and sometimes colour.

Classers are also responsible for ensuring that a pre-shearing check is made to ensure that the wool and sheep areas are free of possible contaminants. They are to supervise shed staff during shearing and train any inexperienced hands. At the end of shearing classers have to provide full documentation concerning the clip.

ROOING

In some primitive sheep (for example in many Shetlands), there is a natural break in the growth of the wool in spring. By late spring this causes the fleece to begin to peel away from the body, and it may then be plucked by hand without cutting – this is known as *rooing*. Individual sheep may reach this stage at slightly different times. Shearing can be done with use of hand-shears or powered shears. Professional sheep shearers can shear a sheep in under a minute, without nicking the sheep.

The fleece is removed in one piece. Second cuts can be made but produce only short fibres, which are more difficult to spin. Primitive breeds, like the Scottish Soay sheep have to be plucked, not sheared, as the kemps are still longer than the soft fleece, (a process called rooing).

SKIRTING

Skirting is disposing of all wool that is unsuitable for spinning. Recovering can be attempted. It can also be done at the same time as carding.

CLEANING

The wool is cleaned. At this point the fleece is full of lanolin and often contains extraneous vegetable matter, such as sticks, twigs, burrs and straw. These may all be removed, though lanolin may be left in the wool till after the spinning, a technique known as spinning 'in the grease'. Indeed if the fabric is to be water repellent, lanolin is not removed at any stage.

Washing the wool at this stage can be a tedious process. Some people wash it a small handful at a time very carefully, and then set it out to dry on a table in the sun. Others will wash the whole fleece. Lanolin is removed by soaking the fleece in very hot water. If the fleece gets agitated, it will become felt, and then spinning is impossible. Felting, when done on purpose (with needles, chemicals, or simply rubbing the fibres against each other), can be used to create garments.

CARDING OR COMBING

It is possible to spin directly from a clean fleece, but it is much easier to spin a carded fleece. Carding by hand yields a rolag, a loose woollen roll of fibres. Using a drum carder yields a bat, which is a mat of fibres in a flat, rectangular shape. Carding mills return the fleece in a roving, which is a stretched bat; it is very long and often the thickness of a wrist.A pencil roving is a roving thinned to the width of a pencil. It can used for knitting without any spinning, or for apprentices.

One good-sized fleece may take weeks to card with a drum-carder, or an eternity by hand. If the fleece is sent to a carding mill, it must be washed before it is carded. Most mills offer washing the wool as a service, with extra fees if the wool is exceptionally dirty. Fibres can be purchased pre-carded. Combing is another method to align the fibres parallel to the yarn, and thus is good for spinning a worsted yarn, whereas the rolag from handcards produces a woolen yarn.

SPINNING

Hand spinning can be done by using a spindle or the spinning wheel. Spinning turns the carded wool fibres into yarn which can then be directly woven, knitted (flat or circular), crocheted, or by other means turned into fabric or a garment. The spinning wheel collects the yarn on a bobbin. A woollen yarn is lightly spun so it is airey, and is a good insulator and suitable for knitting, while a worsted yarn is spun tight to exclude air, and has greater strength and is suited to weaving. Once the bobbin is full, the spinner either puts on a new bobbin, or forms a skein, or balls the yarn. A skein is a coil of yarn twisted into a loose knot. Yarn is skeined using a niddy-noddy or other type of skein -winder. Yarn is rarely balled directly after spinning, it will be stored in skein form, and transferred to a ball only if needed. Knitting from a skein, is difficult as the yarn forms knots, in this case it is best to ball. Yarn to be plied is left on the bobbin.

A skein is either formed on a niddy noddy or some other type of skein winder. Traditionally niddy-noddys looked like an uppercase "i", with the bottom half rotated 90 degrees. Now spinning wheel manufactures also make niddy-noddys that attach onto the spinning wheel for faster skein winding.

PLYING

Plying yarn is when one takes a strand of spun yarn (one strand is often called a single) and spins it together with other strands in order to make a thicker yarn.

Regular plying consists of taking two or more singles and twisting them together, the against their twist. This can be done on a spinning wheel or on a spindle. If the yarn was spun clockwise (which is called a "Z" twist), to ply, the wheel must spin counter-clockwise (an "S" twist). This is the most common way. When plying from bobbins a device called a lazy kate is often used to hold them.

Most spinners (who use spinning wheels) ply from bobbins. This is easier than plying from balls because there is less chance for the yarn to become tangled and knotted if it is simply unwound from the bobbins. So that the bobbins can unwind freely, they are put in a device called a lazy kate, or sometimes simply *kate*.

The simplest lazy kate consists of wooden bars with a metal rod running between them. Most hold between three and four bobbins. The bobbin sits on the metal rod. Other lazy kates are built with devices that create an adjustable amount of tension, so that if the yarn is jerked, a whole bunch of yarn is not wound off, then wound up again in the opposite direction. Some spinning wheels come with a built in lazy kate. Navajo plying consists of making large loops, similar to crocheting.A loop about 8 inches long is made on the leader the end on the leader.

(A leader is the string left on the bobbin to spin off.) The three strands together are spun in the opposite direction. When a third of the loop remains, a new loop is created and the spinning continues.

The process is repeated until the yarn is all plied. The advantage of this method is that only one single is needed and if the single is already dyed this technique allows it to be plied without ruining the colour scheme. This technique also allows the spinner to try to match up thick and thin spots in the yarn, thus making for a smoother end product.

WASHING

If the lanolin is unwanted, and has not already been washed out, this is done now. The skein is tied in six points and steeped overnight in detergent, it is rinsed and air-dried, and re-skinned.

unless the lanolin is to be left in the cloth as a water repellent. When washing a skein it works well to let the wool soak in soapy water overnight, and rinse the soap out in the morning. Dishwashing detergents are commonly used, and a special laundry detergent designed for washing wool is not required.

The dishwashing detergent works and does not harm the wool. After washing, let the wool dry (air drying works best). Once it is dry, or just a bit damp, one can stretch it out a bit on a niddy-noddy.

Putting the wool back on the niddy-noddy makes for a nicer looking finished skein. Before taking a skein and washing it, the skein must be tied up loosely in about six places. If the skein is not tied up, it will be very hard to unravel when done washing.

FLAX

The preparations for spinning is similar across most plant fibres, including Flax and Hemp. Flax is the fibre used to create linen. Cotton is handled differently since it uses the fruit of the plant and not the stalk.

HARVESTING

Flax is pulled out of the ground about a month after the initial blooming when the lower part of the plant begins to turn yellow, and when the most forward of the seeds are found in a soft state. It is pulled in handfuls and several handfuls are tied together with slip knot into a 'beet'. The string is tightened as the stalks dry. The seed heads are removed and the seeds collected, by threshing and winnowing.

RETTING

Threshing and dressing flax at the Roscheider Hof Open Air Museum Retting is the process of rotting away the inner stalk, leaving the outer fibres intact. A standing pool of warm water is needed, into which the beets are submerged. An acid is produced when retting, and it would corrode a metal container. At 80 °F (27 °C), the retting process takes 4 or 5 days, it takes longer takes longer when colder. When the retting is complete the bundles feel soft and slimy, The process can be overdone, and the fibres rot too.

DRESSING THE FLAX

Dressing is removing the fibres from the straw and cleaning it enough to be spun. The flax is broken, scutched and hackled in this step. Breaking The process of breaking breaks up the straw into short segments. The beets are untied and fed between the beater of the breaking machine, the set of wooden blades that mesh together when the upper jaw is lowered.

Scutching In order to remove some of the straw from the fibre a wooden scutching knife is scaped down the fibres while they hang vertically. Heckling Fibre is pulled through various different sized heckling combs. A Heckling comb is a bed of sharp, long-tapered, tempered, polished steel pins driven into wooden blocks at regular spacing. A good progression is from 4 pins per square inch, to 12, to 25 to 48 to 80.

The first three will remove the straw, and the last two will split and polish the fibres. Some of the finer stuff that comes off in the last heckles can be carded like wool and spun. It will produce a coarser yarn than the fibres pulled through the heckles because it will still contain some straw.

SPINNING THE FLAX

Flax can either be spun from a distaff, or from the spinner's lap. Spinners keep their fingers wet when spinning, to prevent forming fuzzy thread. Usually singles are spun with an "S" twist. After flax is spun it is washed in a pot of boiling water for a couple of hours to set the twist and reduce fuzziness.

Many handspinners, will buy a roving of flax. This roving is spun in the same manner as above. The rovings may come with very long fibres (4 to 8 inches), or much shorter fibres (2 to 3 inches).

SILK

Silk is a natural protein fibre, some forms of which can be woven into textiles. The best-known type of silk is obtained from the cocoons of the larvae of the mulberry silkworm *Bombyx mori* reared in captivity (sericulture). The shimmering appearance of silk is due to the triangular prism-like structure of the silk fibre, which allows silk cloth to refract incoming light at different angles, thus producing different colours.

Silks are produced by several other insects, but only the silk of moth caterpillars has been used for textile manufacturing. There has been some research into other silks, which differ at the molecular level. Silks are mainly produced by the larvae of insects undergoing complete metamorphosis, but also by some adult insects such as webspinners. Silk production is especially common in the Hymenoptera (bees, wasps, and ants), and is sometimes used in nest construction. Other types of arthropod produce silk, most notably various arachnids such as spiders (see spider silk).

HISTORY

Woven silk textile from tomb no 1. at Mawangdui in Changsha, Hunan province, China, from the Western Han Dynasty, 2nd century BC

WILD SILK

A variety of wild silks, produced by caterpillars other than the mulberry silkworm have been known and used in China, South Asia, and Europe since ancient times. However, the scale of production was always far smaller than that of cultivated silks. They differ from the domesticated varieties in colour and texture, and cocoons gathered in the wild usually have been damaged by the emerging moth before the cocoons are gathered, so the silk thread that makes up the cocoon has been torn into shorter lengths.

Commercially reared silkworm pupae are killed by dipping them in boiling water before the adult moths emerge, or by piercing them with a needle, allowing the whole cocoon to be unravelled as one continuous thread. This permits a much stronger cloth to be woven from the silk. Wild silks also tend to be more difficult to dye than silk from the cultivated silkworm.

CHINA

Silk fabric was first developed in ancient China, with some of the earliest examples found as early as 3500 BC. Legend gives credit for developing silk to a Chinese empress, Lei Zu (Hsi-Ling-Shih, Lei-Tzu). Silks were originally reserved for the Kings of China for their own use and gifts to others, but spread gradually through Chinese culture and trade both geographically and socially, and then to many regions of Asia.

Silk rapidly became a popular luxury fabric in the many areas accessible to Chinese merchants because of its texture and luster. Silk was in great demand, and became a staple of pre-industrial international trade. In July 2007, archeologists discovered intricately woven and dyed silk textiles in a tomb in Jiangxi province, dated to the Eastern Zhou Dynasty roughly 2,500 years ago.

Although historians have suspected a long history of a formative textile industry in ancient China, this find of silk textiles employing "complicated techniques" of weaving and dyeing provides direct and concrete evidence for silks dating before the Mawangdui-discovery and other silks dating to the Han Dynasty (202 BC-220 AD).

The first evidence of the silk trade is the finding of silk in the hair of an Egyptian mummy of the 21st dynasty, c.1070 BC. Ultimately the silk trade reached as far as the Indian subcontinent, the Middle East, Europe, and North Africa. This trade was so extensive that the major set of trade routes between Europe and Asia has become known as the Silk Road. The highest development was in China.

The Emperors of China strove to keep knowledge of sericulture secret to maintain the Chinese monopoly. Non-etheless sericulture reached Korea around 200 BC, about the first half of the 1st century AD had reached ancient Khotan, and by AD 300 the practice had been established in India.

THAILAND

Silk is produced, year round, in Thailand by two types of silkworms, the cultured Bombycidae and wild Saturniidae. Most production is after the rice harvest in the southern and northeast parts of the country. Women traditionally weave silk on hand looms, and pass the skill on to their daughters as weaving is considered to be a sign of maturity and eligibility for marriage. Thai silk textiles often use complicated patterns in various colours and styles. Most regions of Thailand have their own typical silks.

A single thread filament is too thin to use on its own so women combine many threads to produce a thicker, usable fibre. They do this by hand-reeling the threads onto a wooden spindle to produce a uniform strand of raw silk. The process takes around 40 hours to produce a half kilogram of Thai silk.

Many local operations use a reeling machine for this task, but some silk threads are still hand-re-eled. The difference is that hand-re-eled threads produce three grades of silk: two fine grades that are ideal for lightweight fabrics, and a thick grade for heavier material.

The silk fabric is soaked in extremely cold water and bleached before dyeing to remove the natural yellow colouring of Thai silk yarn. To do this, skeins of silk thread are immersed in large tubs of hydrogen peroxide. Once washed and dried, the silk is woven on a traditional hand operated loom.

INDIA

Silk, known as "Paat" in Eastern India, *Pattu* in southern parts of India and *Resham* in Hindi/Urdu, has a long history in India. Recent archaeological discoveries in Harappa and Chanhu-daro suggest that sericulture, employing wild silk threads from native silkworm species, existed in South Asia during the time of the Indus Valley Civilization, roughly contemporaneous with the earliest known silk use in China. Silk is widely produced today.

India is the second largest producer of silk after China. A majority of the silk in India is produced in Karnataka State, particularly in Mysore and the North Bangalore regions of Muddenahalli, Kanivenarayanapura, and Doddaballapur. India is also the largest consumer of silk in the world. The tradition of wearing silk sarees in marriages by the brides is followed in southern parts of India. Silk is worn by people as a symbol of royalty while attending functions and during festivals.

Historically silk was used by the upper classes, while cotton was used by the poorer classes. Today silk is mainly produced in Bhoodhan Pochampally (also known as Silk City), Kanchipuram, Dharmavaram, Mysore, etc. in South India and Banaras in the North for manufacturing garments and sarees. "Murshidabad silk", famous from historical times, is mainly produced in Malda and Murshidabad district of West Bengal and woven with hand looms in Birbhum and Murshidabad district. Another place famous for production of silk is Bhagalpur.

The silk from Pochampally is particularly well-known for its classic designs and enduring quality. The silk is traditionally hand-woven and hand-dyed and usually also has silver threads woven into the cloth. Most of this silk is used to make sarees. The sarees usually are very expensive and vibrant in colour. Garments made from silk form an integral part of Indian weddings and other celebrations.

In the northeastern state of Assam, three different types of silk are produced, collectively called Assam silk: Muga, Eri and Pat silk. Muga, the golden silk, and Eri are produced by silkworms that are native only to Assam. The heritage of silk rearing and weaving is very old and continues today especially with the production of Muga and Pat *riha* and *mekhela chador*, the three-piece silk sarees woven with traditional motifs. *Mysore Silk Sarees*, which are known for their soft texture, last many years if carefully maintained.

ANCIENT MEDITERRANEAN

In the Odyssey, 19.233, when Odysseus, while pretending to be someone else, is questioned by Penelope about her husband's clothing, he says that he wore a shirt "gleaming like the skin of a dried onion" (varies with translations, literal translation here) which could refer to the lustrous quality of silk fabric. The Roman Empire knew of and traded in silk. During the reign of emperor Tiberius, sumptuary laws were passed that forbade men from wearing silk garments, but these proved ineffectual.

Despite the popularity of silk, the secret of silk-making only reached Europe around AD 550, via the Byzantine Empire. Legend has it those monks working for the emperor Justinian I smuggled silkworm eggs to Constantinople in hollow canes from China. All top-quality looms and weavers were located inside the Palace complex in Constantinople and the cloth produced was used in imperial robes or in diplomacy, as gifts to foreign dignitaries. The remainder was sold at very high prices.

MIDDLE EAST

In Islamic teachings, Muslim men are forbidden to wear silk. Many religious jurists believe the reasoning behind the prohibition lies in avoiding clothing for men that can be considered feminine or extravagant. There are disputes regarding the amount of silk a fabric can consist of (*i.e.*, whether a small decorative silk piece on a cotton caftan is permissible or not) for it to be lawful for men to wear but the dominant opinion of most Muslim scholars is that the wearing of silk for men is forbidden.

Despite injunctions against silk for men, silk has retained its popularity in the Islamic world because of its permissibility for women. The Muslim Moors brought silk with them to Spain during their conquest of the Iberian Peninsula.

MEDIEVAL AND MODERN EUROPE

Venetian merchants traded extensively in silk and encouraged silk growers to settle in Italy. By the 13th century, Italian silk was a significant source of trade. Since that period, the silk worked in the province of Como has been the most valuable silk in the world. The wealth of Florence was largely built on textiles, both wool and silk and other cities like Lucca also grew rich on the trade.

Italian silk was so popular in Europe that Francis I of France invited Armenian silk makers to France to create a French silk industry, especially in Lyon. Mass emigration (especially of Huguenots) during periods of religious dispute had seriously damaged French industry and introduced these various textile industries, including silk, to other countries.

James I attempted to establish silk production in England, purchasing and planting 100,000 mulberry trees, some on land adjacent to Hampton Court Palace, but they were of a species unsuited to the silk worms, and the attempt failed. British enterprise also established silk filature in Cyprus in 1928. In England in the mid 20th Century, silk was produced at Lullingstone Castle in

Kent. Silkworms were raised and reeled under the direction of Zoe Lady Hart Dyke. Production started elsewhere later. In Italy, the Stazione Bacologica Sperimentale was founded in Padua in 1871 to research sericulture. In the late 19th century, China, Japan, and Italy were the major producers of silk.

The most important cities for silk production in Italy were Como and Meldola (Forlì). In medieval times, it was common for silk to be used to make elaborate casings for bananas and other fruits. Silk was expensive in Medieval Europe and used only by the rich. Italian merchants like Giovanni Arnolfini became hugely wealthy trading it to the Courts of Northern Europe.

NORTH AMERICA

James I of England introduced silk-growing to the American colonies around 1619, ostensibly to discourage tobacco planting. The Shakers in Kentucky adopted the practice as did a cottage industry in New England. In the 19th century a new attempt at a silk industry began with European-born workers in Paterson, New Jersey, and the city became a US silk center, although Japanese imports were still more important.

World War II interrupted the silk trade from Japan. Silk prices increased dramatically, and US industry began to look for substitutes, which led to the use of synthetics such as nylon. Synthetic silks have also been made from local, a type of cellulose fibre, and are often difficult to distinguish from real silk (see spider silk for more on synthetic silks).

PROPERTIES

Physical Properties

Silk fibres from the *Bombyx mori* silkworm have a triangular cross section with rounded corners, 5-10 ìm wide. The fibroin-heavy chain is composed mostly of beta-sheets, due to a 59-mer amino acid repeat sequence with some variations. The flat surfaces of the fibrils reflect light at many angles, giving silk a natural shine. The cross-section from other silkworms can vary in shape and diameter: crescent-like for *Anaphe* and elongated wedge for *tussah*.

Silkworm fibres are naturally extruded from two silkworm glands as a pair of primary filaments (brin), which are stuck together, with sericin proteins that act like glue, to form a bave. Bave diameters for tussah silk can reach 65 ìm. See cited reference for cross-sectional SEM photographs.

Silk has a smooth, soft texture that is not slippery, unlike many synthetic fibres. Silk is one of the strongest natural fibres but loses up to 20per cent of its strength when wet. It has a good moisture regain of 11per cent. Its elasticity is moderate to poor: if elongated even a small amount, it remains stretched. It can be weakened if exposed to too much sunlight. It may also be attacked by insects, especially if left dirty.

Silk is a poor conductor of electricity and thus susceptible to static cling. Unwashed silk chiffon may shrink up to 8per cent due to a relaxation of the fibre macrostructure. So silk should either be pre-washed prior to garment construction, or dry cleaned. Dry cleaning may still shrink the chiffon up to 4per cent. Occasionally, this shrinkage can be reversed by a gentle steaming with a press cloth. There is almost no gradual shrinkage nor shrinkage due to molecular-level deformation.

Natural and synthetic silk is known to manifest piezoelectric properties in proteins, probably due to its molecular structure.Silkworm silk was used as the standard for the denier, a measurement of linear density in fibres. Silkworm silk therefore has a linear density of approximately 1 den, or 1.1 dtex.

Comparison ofLinear Density(dtex)	Diameter (μm)	Coeff. Variation	Silk Fibres
Moth: *Bombyx mori*	1.17	12.9	24.8per cent
Spider: *Argiope aurentia*	0.14	3.57	14.8per cent

Chemical properties

Silk emitted by the silkworm consists of two main proteins, sericin and fibroin, fibroin being the structural center of the silk, and serecin being the sticky material surrounding it. Fibroin is made up of the amino acids Gly-Ser-Gly-Ala-Gly-Ala and forms beta pleated sheets. Hydrogen bonds form between chains, and side chains form above and below the plane of the hydrogen bond network.

The high proportion (50per cent) of glycine, which is a small amino acid, allows tight packing and the fibres are strong and resistant to stretching. The tensile strength is due to the many interseeded hydrogen bonds. Since the protein forms a beta sheet, when stretched the force is applied to these strong bonds and they do not break.

Silk is resistant to most mineral acids, except for sulfuric acid, which dissolves it. It is yellowed by perspiration.

USES

Silk's absorbency makes it comfortable to wear in warm weather and while active. Its low conductivity keeps warm air close to the skin during cold weather. It is often used for clothing such as shirts, ties, blouses, formal dresses, high fashion clothes, lingerie, pyjamas, robes, dress suits, sun dresses and kimonos.

Silk's attractive luster and drape makes it suitable for many furnishing applications. It is used for upholstery, wall coverings, window treatments (if blended with another fibre), rugs, bedding and wall hangings

While on the decline now, due to artificial fibres, silk has had many industrial and commercial uses; parachutes, bicycle tires, comforter filling and artillery gunpowder bags. A special manufacturing process removes the outer irritant sericin coating of the silk, which makes it suitable as non-absorbable surgical sutures.

This process has also recently led to the introduction of specialist silk underclothing for children and adults with eczema where it can significantly reduce itch.

PRODUCTION

The cultivation of silk is called sericulture. Over 30 countries produce silk, the major ones are China (54per cent) and India (14per cent).

To produce 1 kg of silk, 104 kg of mulberry leaves must be eaten by 3000 silkworms. It takes about 5000 silkworms to make a pure silk kimono.

Table. Top Ten Cocoons (Re-elable) Producers — 2005

Country	Production (Int $1000)	Footnote	Production (1000 kg)	Footnote
People's Republic of China	978,013	C	290,003	F
India	259,679	C	77,000	F
Uzbekistan	57,332	C	17,000	F
Brazil	37,097	C	11,000	F
Iran	20,235	C	6,000	F
Thailand	16,862	C	5,000	F
Vietnam	10,117	C	3,000	F
Democratic People's Republic of Korea	5,059	C	1,500	F
Romania	3,372	C	1,000	F
Japan	2,023	C	600	F

Note:

No symbol = official figure

F = FAO estimate

* = Unofficial figure

C = Calculated figure

Production in Int $1000 have been calculated based on 1999-2001 international pricesSource: Food And Agricultural Organization of United Nations: Economic And Social Department: The Statistical Division

CULTIVATION

Silk moths lay eggs on specially prepared paper. The eggs hatch and the caterpillars (silkworms) are fed fresh mulberry leaves. After about 35 days and 4 moltings, the caterpillars are 10,000 times heavier than when hatched and are ready to begin spinning a cocoon.

A straw frame is placed over the tray of caterpillars, and each caterpillar begins spinning a cocoon by moving its head in a pattern.

Two glands produce liquid silk and force it through openings in the head called spinnerets. Liquid silk is coated in sericin, a water-soluble protective gum, and solidifies on contact with the air. Within 2–3 days, the caterpillar spins about 1 mile of filament and is completely encased in a cocoon. The silk farmers then kill most caterpillars by heat, leaving some to metamorphose into moths to breed the next generation of caterpillars.

Harvested cocoons are then soaked in boiling water to soften the sericin holding the silk fibres together in a cocoon shape. The fibres are then unwound to produce a continuous thread. Since a single thread is too fine and fragile for commercial use, anywhere from three to ten strands are spun together to form a single thread of silk.

ANIMAL RIGHTS

As the process of harvesting the silk from the cocoon kills the larvae, sericulture has been criticized in the early 21st century by animal rights activists, especially since artificial silks are available. Mohandas Gandhi was also critical of silk production based on the Ahimsa philosophy "not to hurt any living thing."

This led to Gandhi's promotion of cotton spinning machines, an example of which can be seen at the Gandhi Institute. He also promoted *Ahimsa silk*, wild silk made from the cocoons of wild and semi-wild silk moths. Ahimsa silk is promoted in parts of Southern India for those who prefer not to wear silk produced by killing silkworms.

COTTON

Cotton is a soft, fluffy staple fibre that grows in a boll, or protective capsule, around the seeds of cotton plants of the genus *Gossypium*. The plant is a shrub native to tropical and subtropical regions around the world, including the Americas, Africa, India, and Pakistan. The fibre most often is spun into yarn or thread and used to make a soft, breathable textile, which is the most widely used natural-fibre cloth in clothing today. The English name derives from the Arabic *(al) qutn*, which began to be used circa 1400. The botanical purpose of cotton fibre is to aid in seed dispersal.

HISTORY

According to the Foods and Nutrition Encyclopaedia, the earliest cultivation of cotton discovered thus far in the Americas occurred in Mexico, some 8,000 years ago. The indigenous species was *Gossypium hirsutum*, which is today the most widely planted species of cotton in the world, constituting about 89.9per cent of all production worldwide. The greatest diversity of wild cotton species is found in Mexico, followed by Australia and Africa.

Cotton was first cultivated in the Old World 7,000 years ago (5th–4th millennia BC), by the inhabitants of the Indus Valley Civilization, which covered a huge swath of the northwestern part of the Indian subcontinent, comprising today parts of eastern Pakistan and northwestern India. The Indus cotton industry was well developed and some methods used in cotton spinning and fabrication continued to be used until the modern industrialization of India. Well before the Common Era, the use of cotton textiles had spread from India to the Mediterranean and beyond.

Greeks and the Arabs were apparently ignorant about cotton until the Wars of Alexander the Great, as his contemporary Megasthenes told Seleucus I Nicator of "there being trees on which wool grows" in "Indica".

According to The Columbia Encyclopaedia, Sixth Edition:

Cotton has been spun, woven, and dyed since prehistoric times. It clothed the people of ancient India, Egypt, and China. Hundreds of years before the Christian era, cotton textiles were woven in India with matchless skill, and their use spread to the Mediterranean countries. In the first century, Arab traders brought fine muslin and calico to Italy and Spain.

The Moors introduced the cultivation of cotton into Spain in the 9th century. Fustians and dimities were woven there and in the 14th century in Venice and Milan, at first with a linen warp. Little cotton cloth was imported to England before the 15th century, although small amounts were obtained chiefly for candlewicks. By the 17th century, the East India Company was bringing rare fabrics from India. Native Americans skillfully spun and wove cotton into fine garments and dyed tapestries. Cotton fabrics found in Peruvian tombs are said to belong to a pre-Inca culture.

In Iran (Persia), the history of cotton dates back to the Achaemenid era (5th century BC); however, there are few sources about the planting of cotton in pre-Islamic Iran. The planting of cotton was common in Merv, Ray and Pars of Iran. In the poems of Persian poets, especially Ferdowsi's Shahname, there are many references to cotton ("panbe" in Persian). Marco Polo (13th century) refers to the major products of Persia, including cotton. John Chardin, a famous French traveller of 17th century, who had visited the Safavid Persia, has approved the vast cotton farms of Persia.

In Peru, cultivation of the indigenous cotton species *Gossypium barbadense* was the backbone of the development of coastal cultures, such as the Norte Chico, Moche and Nazca. Cotton was grown upriver, made into nets and traded with fishing villages along the coast for large supplies of fish. The Spanish who came to Mexico and Peru in the early 16th century found the people growing cotton and wearing clothing made of it.

During the late medieval period, cotton became known as an imported fibre in northern Europe, without any knowledge of how it was derived, other than that it was a plant; noting its similarities to wool, people in the region could only imagine that cotton must be produced by plant-borne sheep. John Mandeville, writing in 1350, stated as fact the now-preposterous belief: "There grew there [India] a wonderful tree which bore tiny lambs on the endes of its branches.

These branches were so pliable that they bent down to allow the lambs to feed when they are hungrie [*sic*]." (See Vegetable Lamb of Tartary.) This aspect is retained in the name for cotton in many European languages, such as German *Baumwolle*, which translates as "tree wool" (*Baum* means "tree"; *Wolle* means "wool"). By the end of the 16th century, cotton was cultivated throughout the warmer regions in Asia and the Americas.

India's cotton-processing sector gradually declined during British expansion in India and the establishment of colonial rule during the late 18th and early 19th centuries. This was largely due to aggressive colonialist mercantile policies of the British East India Company, which made cotton processing and manufacturing workshops in India uncompetitive. Indian markets were increasingly forced to supply only raw cotton and were forced, by British-imposed law, to purchase manufactured textiles from Britain.

INDUSTRIAL REVOLUTION IN BRITAIN

The advent of the Industrial Revolution in Britain provided a great boost to cotton manufacture, as textiles emerged as Britain's leading export. In 1738, Lewis Paul and John Wyatt, of Birmingham, England, patented the roller spinning machine, and the flyer-and-bobbin system for drawing cotton to a more even thickness using two sets of rollers that traveled at different speeds.

Later, the invention of the spinning jenny in 1764 and Richard Arkwright's spinning frame (based on the roller spinning machine) in 1769 enabled British weavers to produce cotton yarn and cloth at much higher rates. From the late 18th century onwards, the British city of Manchester acquired the nickname *"Cottonopolis"* due to the cotton industry's omnipresence within the city, and Manchester's role as the heart of the global cotton trade.

Production capacity in Britain and the United States was further improved by the invention of the cotton gin by the American Eli Whitney in 1793. Improving technology and increasing control of world markets allowed British traders to develop a commercial chain in which raw cotton fibres were (at first) purchased from colonial plantations, processed into cotton cloth in the mills of Lancashire, and then exported on British ships to captive colonial markets in West Africa, India, and China (via Shanghai and Hong Kong). By the 1840s, India was no longer capable of supplying the vast quantities of cotton fibres needed by mechanized British factories, while shipping bulky, low-price cotton from India to Britain was time-consuming and expensive.

This, coupled with the emergence of American cotton as a superior type (due to the longer, stronger fibres of the two domesticated native American species, *Gossypium hirsutum* and *Gossypium barbadense*), encouraged British traders to purchase cotton from plantations in the United States and the Caribbean. By the mid 19th century, "King Cotton" had become the backbone of the southern American economy. In the United States, cultivating and harvesting cotton became the leading occupation of slaves. During the American Civil War, American cotton exports slumped due to a Union blockade on Southern ports, also because of a strategic decision by the Confederate government to cut exports, hoping to force Britain to recognize the Confederacy or enter the war, prompting the main purchasers of cotton, Britain and France to turn to Egyptian cotton.

British and French traders invested heavily in cotton plantations and the Egyptian government of Viceroy Isma'il took out substantial loans from European bankers and stock exchanges. After the American Civil War ended in 1865, British and French traders abandoned Egyptian cotton and returned to cheap American exports, sending Egypt into a deficit spiral that led to the country declaring bankruptcy in 1876, a key factor behind Egypt's annexation by the British Empire in 1882. During this time, cotton cultivation in the British Empire, especially India, greatly increased to replace the lost production of the American South. Through tariffs and other restrictions, the British government discouraged the production of cotton cloth in India; rather, the raw fibre was sent to England for processing.

The Indian patriot Mahatma Gandhi described the process:

- English people buy Indian cotton in the field, picked by Indian labour at seven cents a day, through an optional monopoly.
- This cotton is shipped on British ships, a three-week journey across the Indian Ocean, down the Red Sea, across the Mediterranean, through Gibraltar, across the Bay of Biscay and the Atlantic Ocean to London. One hundred per cent profit on this freight is regarded as small.
- The cotton is turned into cloth in Lancashire. You pay shilling wages instead of Indian pennies to your workers. The English worker not only has the advantage of better wages, but the steel companies of England get the profit of building the factories and machines. Wages; profits; all these are spent in England.
- The finished product is sent back to India at European shipping rates, once again on British ships. The captains, officers, sailors of these ships, whose wages must be paid, are English. The only Indians who profit are a few lascars who do the dirty work on the boats for a few cents a day.
- The cloth is finally sold back to the kings and landlords of India who got the money to buy this expensive cloth out of the poor peasants of India who worked at seven cents a day. (Fisher 1932 pp 154–156)

In the United States, Southern cotton provided capital for the continuing development of the North. The cotton produced by enslaved African Americans not only helped the South, but also enriched Northern merchants. Much of the Southern cotton was transshipped through the northern ports.

Cotton remained a key crop in the Southern economy after emancipation and the end of the Civil War in 1865. Across the South, sharecropping evolved, in which free black farmers and landless white farmers worked on white-owned cotton plantations of the wealthy in return for a share of the profits. Cotton plantations required vast labour forces to hand-pick cotton, and it was not until the 1950s that reliable harvesting machinery was introduced into the South (prior to this, cotton-harvesting machinery had been too clumsy to pick cotton without shredding the fibres).

During the early 20th century, employment in the cotton industry fell, as machines began to replace labourers, and the South's rural labour force dwindled during the First and Second World Wars. Today, cotton remains a major export of the southern United States, and a majority of the world's annual cotton crop is of the long-staple American variety.

TANGUIS COTTON

In 1901, Peru's cotton industry suffered because of a fungus plague caused by a plant disease known as "cotton wilt" or, more correctly, "fusarium wilt", caused by the fungus *Fusarium vasinfectum*.

The plant disease, which spread throughout Peru, entered plant's roots and worked its way up the stem until the plant was completely dried up. Fermín Tangüis, a Puerto Rican agriculturist who lived in Peru, studied some species of the plant that were affected by the disease to a lesser extent and experimented in germination with the seeds of various cotton plants.

In 1911, after 10 years of experimenting and failures, Tangüis was able to develop a seed which produced a superior cotton plant resistant to the disease. The seeds produced a plant that had a 40per cent longer (between 29 mm and 33 mm) and thicker fibre that did not break easily and required little water. The Tangüis cotton, as it became known, is the variety which is preferred by the Peruvian national textile industry. It constituted 75per cent of all the Peruvian cotton production, both for domestic use and apparel exports. The Tangüis cotton crop was estimated at 225,000 bales that year.

CULTIVATION

Successful cultivation of cotton requires a long frost-free period, plenty of sunshine, and a moderate rainfall, usually from 600 to 1200 mm (24 to 48 inches). Soils usually need to be fairly heavy, although the level of nutrients does not need to be exceptional. In general, these conditions are met within the seasonally dry tropics and subtropics in the Northern and Southern hemispheres, but a large proportion of the cotton grown today is cultivated in areas with less rainfall that obtain the water from irrigation.

Production of the crop for a given year usually starts soon after harvesting the preceding autumn. Planting time in spring in the Northern hemisphere varies from the beginning of February to the beginning of June. The area of the United States known as the South Plains is the largest contiguous cotton-growing region in the world. While dryland (non-irrigated) cotton is successfully grown in this region, consistent yields are only produced with heavy reliance on irrigation water drawn from the Ogallala Aquifer.

Since cotton is somewhat salt and drought tolerant, this makes it an attractive crop for arid and semiarid regions. As water resources get tighter around the world, economies that rely on it face difficulties and conflict, as well as potential environmental problems. For example, improper cropping and irrigation practices have led to desertification in areas of Uzbekistan, where cotton is a major export. In the days of the Soviet Union, the Aral Sea was tapped for agricultural irrigation, largely of cotton, and now salination is widespread.

GENETIC MODIFICATION

Genetically modified (GM) cotton was developed to reduce the heavy reliance on pesticides. The bacterium *Bacillus thuringiensis* (Bt) naturally produces a chemical harmful only to a small fraction of insects, most notably the larvae of moths and butterflies, beetles, and flies, and harmless to other forms of life. The gene coding for BT toxin has been inserted into cotton, causing cotton to produce this natural insecticide in its tissues. In many regions, the main pests in commercial cotton are lepidopteron larvae, which are killed by the BT protein in the transgenic cotton they eat. This eliminates the need to use large amounts of broad-spectrum insecticides to kill lepidopteron pests (some of which have developed pyrethroid resistance). This spares natural insect predators in the farm ecology and further contributes to no insecticide pest management.

BT cotton is ineffective against many cotton pests, however, such as plant bugs, stink bugs, and aphids; depending on circumstances it may still be desirable to use insecticides against these. A 2006 study done by Cornell researchers, the Center for Chinese Agricultural Policy and the Chinese Academy of Science on Bt cotton farming in China found that after seven years these secondary pests that were normally controlled by pesticszide had increased, necessitating the use of pesticides at similar levels to non-Bt cotton and causing less profit for farmers because of the extra expense of GM seeds.

However a more recent 2009 study by the Chinese Academy of Sciences, Stanford University and Rutgers University refutes this. They concluded that the GM cotton effectively controlled bollworm. The secondary pests were mostly miridae (plant bugs) whose increase was related to local temperature and rainfall and only continued to increase in half the villages studied. Moreover, the increase in insecticide use for the control of these secondary insects was far smaller than the reduction in total insecticide use due to BT cotton adoption.

The International Service for the Acquisition of Agri-biotech Applications (ISAAA) said that, worldwide, GM cotton was planted on an area of 16 million hectares in 2009. This was 49per cent of the worldwide total area planted in cotton. The U.S. cotton crop was 93per cent GM in 2010 and the Chinese cotton crop was 68per cent GM in 2009. The initial introduction of GM cotton proved to be a huge success in Australia - the yields were equivalent to the no transgenic varieties and the crop used much less pesticide to produce (85per cent reduction). The subsequent introduction of a second variety of GM cotton led to increases in GM cotton production until 95per cent of the Australian cotton crop was GM in 2009.

The initial introduction of GM cotton proved to be a huge success in Australia - the yields were equivalent to the no transgenic varieties and the crop used much less pesticide to produce (85per cent reduction). The subsequent introduction of a second variety of GM cotton led to increases in GM cotton production until 95per cent of the Australian cotton crop was GM in 2009.

Cotton has also been genetically modified for resistance to glyphosate (marketed as Roundup in North America), an inexpensive and highly effective, but broad-spectrum herbicide. Originally, it was only possible to achieve glyph sate resistance when the plant was young, but with the development of Roundup Ready Flex, it is possible to achieve glyphosate resistance much later in the growing season.

GM cotton acreage in India continues to grow at a rapid rate, increasing from 50,000 hectares in 2002 to 8.4 million hectares in 2009. The total cotton area in India was 9.6 million hectares (the largest in the world or, about 35per cent of world cotton area), so GM cotton was grown on 87per cent of the cotton area in 2009. This makes India the country with the largest area of GM cotton in the world, surpassing China (3.7 million hectares in 2009).

The major reasons for this increase is a combination of increased farm income ($225/ha) and a reduction in pesticide use to control the cotton bollworm. Cotton has gossypol, a toxin that makes it inedible. However, scientists have silenced the gene that produces the toxin, making it a potential food crop.

ORGANIC PRODUCTION

Organic cotton is generally understood as cotton, from plants not genetically modified, that is certified to be grown without the use of any synthetic agricultural chemicals, such as fertilizers or pesticides. Its production also promotes and enhances biodiversity and biological cycles. United States cotton plantations are required to enforce the National Organic Programme (NOP). This institution determines the allowed practices for pest control, growing, fertilizing, and handling of organic crops.As of 2007, 265,517 bales of organic cotton were produced in 24 countries, and worldwide production was growing at a rate of more than 50per cent per year.

PESTS AND WEEDS

The cotton industry relies heavily on chemicals, such as fertilizers and insecticides, although a very small number of farmers are moving towards an organic model of production, and organic cotton products are now available for purchase at limited locations. These are popular for baby clothes and diapers. Under most definitions, organic products do not use genetic engineering.

Historically, in North America, one of the most economically destructive pests in cotton production has been the boll weevil. Due to the US Department of Agriculture's highly successful Boll Weevil Eradication Programme (BWEP), this pest has been eliminated from cotton in most of the United States. This programme, along with the introduction of genetically engineered Bt cotton (which contains a bacterial gene that codes for a plant-produced protein that is toxic to a number of pests such as cotton bollworm and pink bollworm), has allowed a reduction in the use of synthetic insecticides.

Other significant global pests of cotton include the pink bollworm, Pectinophora gossypiella; the chili thrips, Scirtothrips dorsalis; and the cotton seed bug, Oxycarenus hyalinipennis.

HARVESTING

Most cotton in the United States, Europe, and Australia is harvested mechanically, either by a cotton picker, a machine that removes the cotton from the boll without damaging the cotton plant, or by a cotton stripper, which strips the entire boll off the plant. Cotton strippers are used in regions where it is too windy to grow picker varieties of cotton, and usually after application of a chemical defoliant or the natural defoliation that occurs after a freeze. Cotton is a perennial crop in the tropics, and without defoliation or freezing, the plant will continue to grow. Cotton continues to be picked by hand in developing countries.

COMPETITION FROM SYNTHETIC FIBRES

The era of manufactured fibres began with the development of rayon in France in the 1890s. Rayon is derived from a natural cellulose and cannot be considered synthetic, but requires extensive processing in a manufacturing process, and led the less expensive replacement of more naturally derived materials. A succession of new synthetic fibres were introduced by the chemicals industry in the following decades. Acetate in fibre form was developed in 1924.

Nylon, the first fibre synthesized entirely from petrochemicals, was introduced as a sewing thread by DuPont in 1936, followed by DuPont's acrylic in 1944. Some garments were created from fabrics based on these fibres, such as women's hosiery from nylon, but it was not until the introduction of polyester into the fibre marketplace in the early 1950s that the market for cotton came under threat.The rapid uptake of polyester garments in the 1960s caused economic hardship in cotton-exporting economies, especially in Central American countries, such as Nicaragua, where cotton production had boomed tenfold between 1950 and 1965 with the advent of cheap chemical pesticides. Cotton production recovered in the 1970s, but crashed to pre-1960 levels in the early 1990s.

Beginning as a self-help programme in the mid-1960s, the Cotton Research and Promotion Programme (CRPP) was organized by U.S. cotton producers in response to cotton's steady decline in market share. At that time, producers voted to set up a per-bale assessment system to fund the programme, with built-in safeguards to protect their investments.

With the passage of the Cotton Research and Promotion Act of 1966, the programme joined forces and began battling synthetic competitors and re-establishing markets for cotton. Today, the success of this programme has made cotton the best-selling fibre in the U.S. and one of the best-selling fibres in the world.

Administered by the Cotton Board and conducted by Cotton Incorporated, the CRPP works to greatly increase the demand for and profitability of cotton through various research and promotion activities. It is funded by U.S. cotton producers and importers.

USES

Cotton is used to make a number of textile products. These include terrycloth for highly absorbent bath towels and robes; denim for blue jeans; chambray, popularly used in the manufacture of blue work shirts (from which we get the term "blue-collar"); and corduroy, seersucker, and cotton twill. Socks, underwear, and most T-shirts are made from cotton. Bed sheets often are made from cotton.

Cotton also is used to make yarn used in crochet and knitting. Fabric also can be made from recycled or recovered cotton that otherwise would be thrown away during the spinning, weaving, or cutting process. While many fabrics are made completely of cotton, some materials blend cotton with other fibres, including rayon and synthetic fibres such as polyester. It can either be used in knitted or woven fabrics, as it can be blended with elastine to make a stretchier thread for knitted fabrics, and apparel such as stretch jeans.

In addition to the textile industry, cotton is used in fishnets, coffee filters, tents, gunpowder (see nitrocellulose), cotton paper, and in bookbinding. The first Chinese paper was made of cotton fibre. Fire hoses were once made of cotton.

The cottonseed which remains after the cotton is ginned is used to produce cottonseed oil, which, after refining, can be consumed by humans like any other vegetable oil.

The cottonseed meal that is left generally is fed to ruminant livestock; the gossypol remaining in the meal is toxic to monogastric animals. Cottonseed hulls can be added to dairy cattle rations for roughage. During the American slavery period, cotton root bark was used in folk remedies as an abortifacient, that is, to induce a miscarriage.

Cotton linters are fine, silky fibres which adhere to the seeds of the cotton plant after ginning. These curly fibres typically are less than 1/8 in (3 mm) long. The term also may apply to the longer textile fibre staple lint as well as the shorter fuzzy fibres from some upland species. Linters are traditionally used in the manufacture of paper and as a raw material in the manufacture of cellulose.

In the UK, linters are referred to as "cotton wool". This can also be a refined product (*absorbent cotton* in U.S. usage) which has medical, cosmetic and many other practical uses. The first medical use of cotton wool was by Dr. Joseph Sampson Gamgee at the Queen's Hospital (later the General Hospital) in Birmingham, England.

Shiny cotton is a processed version of the fibre that can be made into cloth resembling satin for shirts and suits.

However, it is hydrophobic (does not absorb water easily), which makes it unfit for use in bath and dish towels (although examples of these made from shiny cotton are seen).

The term Egyptian cotton refers to the extra long staple cotton grown in Egypt and favoured for the luxury and upmarket brands worldwide. During the U.S. Civil War, with heavy European investments, Egyptian-grown cotton became a major alternate source for British textile mills. Egyptian cotton is more durable and softer than American Pima cotton, which is why it is more expensive. Pima cotton is American cotton that is grown in the southwestern states of the U.S.

INTERNATIONAL TRADE

The largest producers of cotton, currently (2009), are China and India, with annual production of about 34 million bales and 24 million bales, respectively; most of this production is consumed by their respective textile industries. The largest exporters of raw cotton are the United States, with sales of $4.9 billion, and Africa, with sales of $2.1 billion. The total international trade is estimated to be $12 billion. Africa's share of the cotton trade has doubled since 1980.

Neither area has a significant domestic textile industry, textile manufacturing having moved to developing nations in Eastern and South Asia such as India and China. In Africa, cotton is grown by numerous small holders. Dunavant Enterprises, based in Memphis, Tennessee, is the leading cotton broker in Africa, with hundreds of purchasing agents. It operates cotton gins in Uganda, Mozambique, and Zambia.

In Zambia, it often offers loans for seed and expenses to the 180,000 small farmers who grow cotton for it, as well as advice on farming methods. Cargill also purchases cotton in Africa for export.

The 25,000 cotton growers in the United States are heavily subsidized at the rate of $2 billion per year.

The future of these subsidies is uncertain and has led to anticipatory expansion of cotton brokers' operations in Africa. Dunavant expanded in Africa by buying out local operations. This is only possible in former British colonies and Mozambique; former French colonies continue to maintain tight monopolies, inherited from their former colonialist masters, on cotton purchases at low fixed prices.

LEADING PRODUCER COUNTRIES

Table. Top Ten Cotton Producers — 2009(480-Pound Bales).

People's Republic of China	32.0 million bales
India	23.5 million bales
United States	12.4 million bales
Pakistan	10.8 million bales
Brazisl	5.5 million bales
Uzbekistan	4.4 million bales
Australia	1.8 million bales
Turkey	1.7 million bales
Turkmenistan	1.1 million bales
Syria	1.0 million bales

The five leading exporters of cotton in 2009 are:

1. The United States,
2. India,
3. Uzbekistan,
4. Pakistan,
5. Brazil.

The largest non-producing importers are Korea, Russia, Taiwan, Japan, and Hong Kong.

In India, the states of Maharashtra (26.63per cent), Gujarat (17.96per cent) and Andhra Pradesh (13.75per cent) and also Madhya Pradesh are the leading cotton producing states, these states have a predominantly tropical wet and dry climate.

In Pakistan, cotton is grown predominantly in the provinces of Punjab and Sindh. The leading city in cotton production is the Punjabi city of Faisalabad which is also leading in textiles within Pakistan. The Punjab has a tropical wet and dry climate throughout the year therefore enhancing the growth of cotton.

In the United States, the state of Texas led in total production as of 2004, while the state of California had the highest yield per acre.

FAIR TRADE

Cotton is an enormously important commodity throughout the world. However, many farmers in developing countries receive a low price for their produce, or find it difficult to compete with developed countries.

This has led to an international dispute (see United States – Brazil cotton dispute):

On 27 September 2002, Brazil requested consultations with the US regarding prohibited and actionable subsidies provided to US producers, users and/or exporters of upland cotton, as well as legislation, regulations, statutory instruments and amendments thereto providing such subsidies (including export credits), grants, and any other assistance to the US producers, users and exporters of upland cotton.

On 8 September 2004, the Panel Report recommended that the United States "withdraw" export credit guarantees and payments to domestic users and exporters, and "take appropriate steps to remove the adverse effects or withdraw" the mandatory price-contingent subsidy measures.

In addition to concerns over subsidies, the cotton industries of some countries are criticized for employing child labour and damaging workers' health by exposure to pesticides used in production. The Environmental Justice Foundation has campaigned against the prevalent use of forced child and adult labour in cotton production in Uzbekistan, the world's third largest cotton exporter.

The international production and trade situation has led to "fair trade" cotton clothing and footwear, joining a rapidly growing market for organic clothing, fair fashion or so-called "ethical fashion".

The fair trade system was initiated in 2005 with producers from Cameroon, Mali and Senegal.

TRADE

Cotton is bought and sold by investors and price speculators as a tradable commodity on 2 different stock exchanges in the United States of America.

- Cotton futures contracts are traded on the New York Mercantile Exchange (NYMEX) under the ticker symbol TT. They are delivered every year in March, May, July, October, and December.
- Cotton #2 futures contracts are traded on the New York Board of Trade (NYBOT) under the ticker symbol CT. They are delivered every year in March, May, July, October, and December.

CRITICAL TEMPERATURES

- *Favourable travel temperature range*: below 25°C (77°F)
- *Optimum travel temperature*: 21°C (70°F)
- *Glow temperature*: 205°C (401°F)
- *Fire point*: 210°C (410°F)
- *Autoignition temperature*: 407°C (765°F)
- *Autoignition temperature (for oily cotton)*: 120°C (248°F)

Cotton dries out, becomes hard and brittle and loses all elasticity at temperatures above 25°C (77°F). Extended exposure to light causes similar problems. A temperature range of 25°C (77°F) to 35°C (95°F) is the optimal range for mold development. At temperatures below 0°C (32°F), rotting of wet cotton stops. Damaged cotton is sometimes stored at these temperatures to prevent further deterioration.

BRITISH STANDARD YARN MEASURES

- 1 thread = 55 inches (about 137 cm)
- 1 skein or rap = 80 threads (120 yards or about 109 m)
- 1 hank = 7 skeins (840 yards or about 768 m)
- 1 spindle = 18 hanks (15,120 yards or about 13.826 km)

FIBRE PROPERTIES

Property	Evaluation
Shape	Fairly uniform in width, 12-20 micrometers; length varies from 1 cm to 6 cm (½ to 2½ inches); typical length is 2.2 cm to 3.3 cm (7/8 to 1¼ inches).
Luster	high
Tenacity (strength)	
Dry	3.0-5.0 g/d
Wet	3.3-6.0 g/d
Resiliency	low
Density	1.54-1.56 g/cm³
Moisture absorption	
raw: conditioned	8.5per cent
saturation	15-25per cent
mercerized: conditioned	8.5-10.3per cent
saturation	15-27per cent+
Dimensional stability	good

The chemical composition of cotton is as follows:

- Cellulose 91.00per cent
- Water 7.85per cent
- Protoplasm, pectins 0.55per cent

- Waxes, fatty substances 0.40per cent
- Mineral salts 0.20per cent

COTTON GENOME

A public genome sequencing effort of cotton was initiated [45] in 2007 by a consortium of public researchers. They agreed on a strategy to sequence the genome of cultivated, tetraploid cotton. "Tetraploid" means that cultivated cotton actually has two separate genomes within its nucleus, referred to as the A and D genomes. The sequencing consortium first agreed to sequence the D-genome relative of cultivated cotton (G. raimondii, a wild Central American cotton species) because of its small size and limited number of repetitive elements.

It is nearly one-third the number of bases of tetraploid cotton (AD), and each chromosome is only present once. The A genome of G. arboreum would be sequenced next. Its genome is roughly twice the size of G. raimondii's. Part of the difference in size between the two genomes is the amplification of retrotransposons (GORGE).

Once both diploid genomes are assembled, then research could begin sequencing the actual genomes of cultivated cotton varieties. This strategy is out of necessity; if one were to sequence the tetraploid genome without model diploid genomes, the euchromatic DNA sequences of the AD genomes would co-assemble and the repetitive elements of AD genomes would assembly independently into A and D sequences respectively. Then there would be no way to untangle the mess of AD sequences without comparing them to their diploid counterparts.

The public sector effort continues with the goal to create a high-quality, draft genome sequence from reads generated by all sources. The public-sector effort has generated Sanger reads of BACs, fosmids, and plasmids as well as 454 reads. These later types of reads will be instrumental in assembling an initial draft of the D genome.

They announced that they would donate their raw reads to the public. This public relations effort gave them some recognition for sequencing the cotton genome. Once the D genome is assembled from all of this raw material, it will undoubtedly assist in the assembly of the AD genomes of cultivated varieties of cotton, but a lot of hard work remains.

YIELDPER CENT MAXIMIZATION IN THE COTTON SPINNING INDUSTRY (YARN MANUFACTURING)

Yieldper cent shows the performance of any industry, it shows efficiency of the industry to convert the raw material into the finished goods. The industry management focuses so much on the maximization of yield so that maximum profit can be earned. Yieldper cent directly relates with profit of the industry, textile cotton spinning industry usually give the yieldper cent up to 84, this can be increased up to 1 to 2per cent, adding million in the profit of cotton spinning industry. This article presents the methodology to enhance the yieldper cent of cotton spinning industry.

INTRODUCTION

Raw material consist impurities up to 4-16per cent(according to type, region and the efficiency of the ginning), the yarn manufacturing processes (Blow-room to carding) are especially designed to remove these impurities. While removing the impurities some good fibres are also removed due to inefficient machine setting or improper machine sequence selection and there are many more causes which decrease the yieldper cent. Some of the causes are discussed here which increase/decrease the yieldper cent such as

- Raw material selection
- Raw martial process route selection
- Improper machine setting
- Maintenance of Machines
- Inefficient air conditioning plant
- Improper material handling

RAW MATERIAL SELECTION

Raw martial costs about 50-70per cent of the total cost of the product[1]. It is mentioned above that impurities in the cotton raw material in Pakistan ranges between 4 to 16 per cent, these impurities must be removed to achieve the required quality product. If the raw material for the yarn is not properly selected than it may cause decrease in the yieldper cent.

Finer yarn requires more cleaning of the material than the coarser yarn because impurities present in the fine yarn create hinders in further processing and can be more visible and vice versa than coarse yarn. Finer yarn required fine quality raw cotton which have sufficient strength and the fibre length if the cotton selected for the finer yarn is of not sufficient length and strength then the good fibres may be wasted during the processing in result of that yieldper cent is decreased.

Another main characteristic of raw material which also contribute in decrease of yieldper cent is extra moisture in material because if the material has extra moister present in it then it may difficulty to separate the impurities and good fibres which required more efforts, energy, time etc such material sticks with machine parts which may go into the waste and become the cause in yieldper cent decrease.

YUCCA FIBRE

Yucca fibres were at one time widely used throughout Central America for many things. Currently they are mainly used to make twine. Yucca leaves are harvested and then cut to a standard size. The leaves are crushed in between two large rollers producing the fibres which are bundled up and dried in the sun over trellises. The dried fibres are combined into rolags. At this point it is ready to spin. The waste, a pulpy liquid that stinks, can be used as a fertilizer.

RAW MATERIAL PROCESS ROUTE SELECTION

Route selection for processing the raw material is also another important task for the management. In cotton spinning industry different lines of processing are installed in the Blow Room, routes have different beating points and according to amount of impurities present in material and quality requirement the processing route is selected. Which reduces the wastage of good fibre, resulting increase in (production) yieldper cent. Processing route selection is based on the type of the material process *e.g.* natural and synthetic material cannot be processed on the same route even you are required to manufacturing the pc (Polyester cotton with any ration) for blend yarn you have to process both fibres separately and at any particular point blend them with required ratio.

Now a days technology facilities process and separate blending machines are available to blend the material with any ratio or you can blend the slivers of the different materials at the draw frame.

Material passage at the draw frame is also pre-decided whether material should be double (Breaker and finisher) or triple passage (breaker, intermediate and finisher). With less passage the production increased by saving the wasteper cent *i.e.* 0.5 but the quality may be effected or not if the raw material quality is enough then the intermediate process may be omitted otherwise It should be carried to improve the quality so that final product can be sold. If after the processing product is not sold then it means whole material and efforts consumed on it are wasted and yieldper cent becomes 0. Therefore it should be kept in mind that product must be sell out.

IMPROPER MACHINE SETTING

Required quality of the product cannot be achieved without the proper setting of machine, any spinning machine setting parameters includes

- Speed
- Gauge
- Top Roller Pressure
- Air pressure

With the above mentioned setting parameters you can manufacture the required quality of yarn within the required time frame. With the speed you can adjust the draft, production rate, beats per unit time, and many more. With the gauge and pressure (Air and top Roller) setting you can easily process the material without deteriorating its quality and properly removing the impurities to achieve the targeted product and the result of this surely contributes in the increase of yieldper cent. If any above said setting parameter is improperly set then it will deteriorate the quality and ultimately decrease the yieldper cent.

MAINTENANCE OF MACHINES:

There is old saying that machine cannot become older, if properly maintained time to time, if machine is not properly maintained then it will not deliver the required quality and may also waste good fibres and due to this yieldper cent will decrease.

E.g. if card machine is not properly overhauled and its taker-in and cylinder wires are out of order, when the material is process on such a machine then the machine is unable to remove the impurities and also unable to properly open the fibres up to individual fibre stage, then the product (sliver) delivered by the card machine is not match with requirement thus for the required quality most of good material may become useless or you have option to reprocess the material in both cases yieldper cent is ultimately decrease and any owner/ management cants bear this, therefore for improving the yieldper cent machine maintenance should be primary objective well maintained machines produces required/targeted quality and yield.

INEFFICIENT AIR CONDITIONING PLANT

Air-conditioning plant contributes most of its share in increase/decrease of the yieldper cent, due to A.C plant the yieldper cent can be at maximum or minimum level therefore if plant is properly maintained providing proper suctions into transportation pipes, machines, atmospheric conditions with the requirement of each department then surely yieldper cent can be increase.

The main objectives of the A.C plant is to maintenance required temperature and RHper cent into every department, provide proper suction into return ducts so that the fluf and good fibres from the floor properly collected and have sufficient air filters, so that the dust, other impurities and good fibres can be separated.

Each machine have also separate suction system *e.g.* at ring frame pnewmafil is collected by the internal machine suction. Pnemafil 100per cent good fibres which and be reprocessed with minimum efforts if these are not collected then went into waste hence causing decrease in yieldper cent. Similarly A.C plant performs its job in every machine to prevent the wastage of good/processable fibre.

THE OBJECTIVES OF THE AIR-CONDITIONING PLANTS INCLUDES

- Control the RHper cent and Temperature in every department
- Collect and Control the fulf
- Properly separate the collected impurities and good fibre by the help of filters

By achieving the above mentioned objectives we can improve the efficiency of man and machine, Quality of yarn and production of plant in result of that we can achieve our target *i.e.* yieldper cent maximization

IMPROPER MATERIAL HANDLING:

Improper handling of material before, during and after process decreases the yieldper cent. *E.g.* the slivers cans during transporting from the card to draw frame or from draw to comber or simplex if not properly carried and by chance material may be deteriorate by hand touching or due to other reason can fall on floor because of uneven floor or cane wheels will contribute in yieldper cent shortage.

Handling of material start when the raw material arrived and you have to plane well where to store it and in which conditions it should me stored considering that when transporting to production floor may not be effected/damage/detoriated.

Material handling is also required in each department of cotton spinning staring from blow-room as the material is process through each department *e.g.* final product of blow-room is lap or final product of chute feed system is sliver of card therefore the lap/slivers are properly transported and stored so that their quality cannot be effected similarly the final product of each department is well stored, material handling also deals with the proper storage and transportation of the finished product.

PHOTOCHROMIC

Photochromism is the reversible transformation of a chemical species between two forms by the absorption of electromagnetic radiation, where the two forms have different absorption spectra.

This can be described as a reversible change of colour upon exposure to light? This property is a boon for scientists doing research on intelligent textiles where they are making use of this property to store data on the surface of textile fabrics and polymer sheets. Whereas the same property of some reactive dyes is a bane for textile processors. The change in shade after dyeing creates unwanted problems in dyeing. An optical recording medium contains, on a base, one or more dyes and a polymer which forms liquid-crystalline phases. The information is written into the uniformly oriented liquid-crystalline polymer layer, for example by means of a laser. During this procedure, the polymer heats up locally to above a phase transition temperature. By cooling, the resulting change is frozen in the glass state.

The information can be erased by applying an electric field and/or heating. The recording material permits high-contrast storage and possesses high sensitivity, good resolution and excellent stability. There are other chromatic properties called electrochromatism and thermochromatism of dyes that are affected by electric field and heat respectively.

Photochromic colours are plastisol-based inks, which are off-white when not exposed to UV radiation. It gains colour when exposed to Sun light/UV light. The colour change is "reversible," *i.e.*, the colour will fade again and appear colour less upon removal from UV light/sun light exposure. These inks are available in various colours. See our colour availability chart for a complete list of available colours.

The uses of Americos Photo chromic colours are:

- On garment to create novel products and promotional items like T-shirts
- On fabric/garment to print company logo/brand name to prevent duplication
- On garment which are used for party wear
- Thermometers and temperature indicators
- Security printing
- Food industry to indicate temperature of packaged food

NANO TEXTILE

WHAT IS NANOTECHNOLOGY?

Nanotechnology is the technical process of working on the nano-scale – each nano-scale molecule is one million times smaller than a grain of sand. Nanotechnology refers to not only the small size of the materials being used, but also how those materials are engineered to perform specific functions. Traditional coatings make garments feel stiff and clog the weave of the fabric preventing breath ability.

Using nanotechnology, our treatments are small enough to attach to individual fibres, delivering superior performance characteristics without compromising the look, feel or comfort of the fabric.

COOLEST COMFORT

Night before the big job interview. Sleep elusive. Hot, then cold, then hot again. No fever, just frantic. Good thing your sheets are keeping you cool. Wake up to sunshine, fresh outlook. Feel like a million bucks. Dress to impress. Blow them away with energy and intelligence. Career happily underway. With Nano Textiles Coolest Comfort, you stay dry and comfortable.

An advanced moisture-wicking system helps balance your body temperature to keep you feeling fresh.

Keep dry, stay cool, rest easy.

- Balances body temperature
- Enhances comfort
- Retains fabric's natural softness
- Allows fabric to breathe naturally

RESISTS SPILLS

Playing hooky. Breakfast in bed. Perfect white linens. Reading the *Times*. Everything just right. Phone rings. Newspaper falls on breakfast tray. Cranberry juice flies on favourite duvet. Relaxed mood is ruined. Good thing your bed cover isn't.

Conventional methods only coat fabrics superficially. But Nano Textiles uses an innovative process to build spill resistance into the individual fibres, so fabrics hold up under the toughest spills. You'll never sacrifice softness or durability, though - Nano Textiles technology makes your duvets, drapery, tablecloths, placemats, and even mattress pads stay looking beautiful, longer.

If it could, your bed would thank you.

- Repels liquids
- Outperforms conventional fabric treatments
- Provides long lasting protection
- Extends the life of the fabric
- Retains fabric's natural softness
- Allows fabric to breathe naturally

REPELS AND RELEASES STAINS

10th annual family reunion. More relatives than you can count. Potluck dinner. Plates piled high. Favourite tablecloth "decorated" with Grandma's special spaghetti sauce, and who knows what else. Luckily, your tablecloth is enhanced with an invisible shield. Nano Textiles Repels and Releases Stains technology actually prevents stains from sinking into fabrics, so spills are never a problem. Easy cleaning and durable protection keeps your finery looking its finest.

- Repels spills
- Helps stains wash out easily
- Provides long lasting protection
- Extends the life of the fabric
- Retains fabric's natural softness
- Allows fabric to breathe naturally

RESISTS STATIC

Cozy evening in. An old movie in front of a roaring fire. Hot popcorn. Great date. Warm blanket- It's not the TV with the bad static, it's the blanket. Dust, lint, and dog hair getting in the way of real electricity. Fast forward. New blanket. Same great date. The right kind of attraction.

Nano Textiles provides permanent static resistance for throws and blankets. It repels hair, lint, and dust while staying soft, comfortable, and beautifully durable. Let the sparks fly, just not between you and your blankets.

- Provides permanent static protection
- Repels lint, dust, dirt and pet hair
- Enhances appearance and comfort
- Retains fabric's natural softness
- Allows fabric to breathe naturally

ORGANIC CLOTHING: WEAR WITH AWARENESS

When picking cotton clothing, today's ecologically minded consumer has several choices. Once believed to be a pure, natural fabric, today's cotton is dosed with harmful chemicals that pollute our environment. Today, thanks to strong media awareness and a revitalization of old techniques, we have alternatives. Organically grown cotton, naturally coloured cotton and recycled cotton products give us three choices for truly natural clothing. It may cost a little more, but by supporting the natural cotton industry's growth, we're supporting ourselves and the future of our planet.

PROBLEMS WITH CONVENTIONAL COTTON

The use of cotton dates back to the Egyptians some 4,500 year ago. For thousands of years, cotton, grown using natural methods, was the primary source of textiles. Today, cotton production is far from natural. The United States alone dumps 8.5 million tons of pesticides on cotton fields annually. And that's not all. Conventionally grown cotton accounts for nearly 25 per cent of the total insecticide use for crops worldwide. This chemical onslaught harms our environment by polluting ground waters and soil, resulting in the death of wildlife and natural habitats. Most of the cotton garments sold today is made from cotton grown with chemical pesticides, bleached and then coloured with chemical dyes containing toxic heavy metals. About a third of a pound of chemicals are used to make one adult T-shirt; two-thirds of a pound of chemicals can go into a pair of jeans.

THREE ECO-FRIENDLY ALTERNATIVES

Concerned consumers are choosing clothing made from organic cotton, naturally coloured cotton, and recycled sources. Here are details on these three environmentally healthy ways of producing cotton clothing and other textiles.

ORGANICALLY GROWN COTTON

Organic cotton is grown without the use of synthetic chemical fertilizers, pesticides or defoliants. By incorporating farming practices that increase fertility and diverse eco-systems, organic farmers rely on time-honoured techniques. Crop rotation, cover crops, organic fertilizers, integrated pest management, and human labour for weed control are a few of the methods used. To be certified organic, the soil must be free of synthetic pesticides for at least three years. Farmers and processors are required to pass yearly inspections.

With the obvious environmental and health benefits of growing cotton organically, why aren't more farmers doing it? In the past, demand for organic cotton has been limited due to higher production costs. However, in recent years media attention has been strong. Some major industry players such as Levi Strauss, Nike and The Gap are blending organic cotton fibres with conventionally produced cotton. The growing public awareness of the toxicity of conventional farming methods has resulted in an increase in both the demand for organically grown cotton and the amount of acreage planted.

NATURALLY COLOURED COTTON

Until recently, chemical dyeing was counted as the only viable way to colour cotton clothing. This process requires several steps, each of which creates toxic waste. Cotton is often bleached before it is dyed and heavy metal mordants are used to adhere the dye to the fabric. Because dyes have a hard time adhering to cotton, at least half of the chemicals end up as waste water in rivers and in the soil. Even in small amounts these heavy metals are lethal.

In 1982, entomologist Sally Fox reintroduced naturally coloured cotton, eliminating the need for dyeing altogether. Cottons of different colours have always existed in nature. Like our eyes or hair, cotton is genetically encoded with colours ranging from brown to tan. Native peoples have used these wild cottons for weaving and hand spinning for centuries. Because of its short fibres and inherent weakness, however, it was unable to be processed by modern textile machinery and had limited commercial value.

With a programme of plant breeding, Fox developed a strong, long-fibre, coloured cotton that can be used commercially. Today, coloured cotton is grown on the stem in shades of brown, reddish brown, green and yellow, totally eliminating the need for dyes.

In the long run, no dyeing means a savings in our pocket. The cost of one pound of dyed cotton including dyestuff, water, energy costs and toxic waste disposal is 20 to 40 per cent higher than that of coloured cotton. And with coloured cotton there are no hidden costs to the environment.

In addition coloured cotton offers:

- An eco-friendly solution. Since there is no bleaching or harmful dyestuff involved in manufacturing, no waste water is produced.
- Unlike dyed material, the colour of the fabrics made from coloured cotton actually deepens with washing.
- Coloured cotton is naturally pest and disease tolerant, making it easier to grow organically.
- Coloured cotton is suitable for chemically sensitive people.
- Coloured cotton provides new yarn and fabric design potentials.
- Without the need for abundant water or energy sources, new mill sites have many more choices for location.

In recent years, the demand for coloured cotton has increased. Large companies like Esprit and Levi have launched popular "green lines" of cotton outerwear using dye-free, unbleached, organic green cotton. The demand for these items far exceed the supply. Coloured cotton is now being grown in the United States, Europe and Australia.

RECYCLED COTTON

Another ecological choice when purchasing cotton clothing is Eco Fibre. Eco Fibre is a recycled cotton fabric made from recovered cotton that would otherwise be cast off during the spinning, weaving or cutting process. There are no harsh chemicals used in the processing of this fabric.

The next time you are shopping for cotton clothing, choose garments made from organic cotton, naturally coloured cotton or recycled cotton. By buying from these natural sources, you are supporting the demand for a new earth-friendly textile industry. Your choice makes a difference.

Hemp is naturally one of the most ecologically friendly fabrics and also the oldest. The Columbia History of the World states that the oldest relics of human industry are bits of hemp fabric discovered in tombs dating back to approximately 8,000 BC.

Hemp fibre is one of the strongest and most durable natural textile fibres. Not only is it strong, but it also holds its shape having one of the lowest per cent elongation of any natural fibre. In fact, its combination of ruggedness and comfort were utilized by Levi Strauss as a lightweight duck canvas for the very first pair of jeans made in California. Furthermore hemp has the best ratio of heat capacity of all fibres giving it superior insulation properties.

As a fabric, hemp provides all the warmth and softness of other natural textiles but with a superior durability seldom found in other materials. Natural organic hemp fibre 'breathes' and is biodegradable. Hemp blended with other fibres easily incorporate the desirable qualities of both textiles. When combined with the natural strength of hemp, the soft elasticity of cotton or the smooth texture of silk create a whole new genre of fashion design.

A fibre of a hundred uses besides fabrics, hemp is also used in the production of paper.

The oldest piece of paper - over 2000 years old - was discovered in China and is made from hemp. Until 1883, between 75per cent and 90per cent of all paper in the world was made with hemp fibre. The Gutenberg bible (15th century), Lewis Carroll's Alice in Wonderland (19th century) and just about everything in between was printed on hemp paper.

Thomas Jefferson wrote the early drafts of the Declaration of Independence on hemp paper produced in Holland. Jefferson grew hemp on his plantation as an industrial crop, selling the dried stalk to the U.S. Navy as outfitting material. George Washington also grew hemp, harvesting the fibrous seed for a variety of commercial uses including a skin lotion.

Other uses include feed for animals and for humans in veggie burgers, salad dressings, and pastas. Hemp seed is nutritious and contains more essential fatty acids than any other source, is second only to soybeans in complete protein (but is more digestible by humans), is high in B-vitamins, and is a good source of dietary fibre. Cosmetics manufacturers include hemp oil in makeup, skin lotions, and shampoo. In Europe, hemp is used in household cleaners as a natural alternative to harsher chemicals.

Hemp is a renewable resource which grows more quickly and easily than trees making hemp more cost effective than waiting decades for trees to grow to be used in man-made fibre production such as lyocell and rayon from wood pulps.

The bark of the hemp stalk contains bast fibres, which are among the Earth's longest natural soft fibres and are also rich in cellulose.

The cellulose and hemi-cellulose in its inner woody core are called hurds. Hemp fibre is longer, stronger, more absorbent and more insulative than cotton fibre.

Hemp produces more pulp per acre than timber on a sustainable basis, and can be used for every quality of paper. Hemp paper manufacturing can reduce wastewater contamination.

Hemp's low lignin content reduces the need for acids used in pulping, and its creamy colour lends itself to environmentally-friendly bleaching instead of harsh chlorine compounds. Less bleaching results in less dioxin and fewer chemical by-products.

Hemp fibre paper resists decomposition, and does not yellow with age when an acid-free process is used.

Hemp paper more than 1,500 years old has been found. Hemp paper can also be recycled more times than wood-based paper.

According to the Department of Energy, hemp is an excellent biomass fuel producer and the hydrocarbons in hemp can be processed into a wide range of biomass energy sources, from fuel pellets to liquid fuels and gas. Development of bio-fuels could significantly reduce our consumption of fossil fuels and nuclear power.

Hemp can be grown organically easily and hemp is most often grown without herbicides, fungicides or pesticides. Hemp is also a natural weed suppressor due to the fast growth of the plant's canopy. Eco-friendly hemp can replace most toxic petrochemical products.

Research is being done to use hemp in manufacturing biodegradable plastic products: plant-based cellophane, recycled plastic mixed with hemp for injection-molded products, and resins made from the oil are just a few examples.

- Properties of Hemp
- Uses for Hemp
- Hemp Fibres and Fabrics
- Manufacturing Hemp Fabric
- Dyeing and Finishing
- Ecology of Hemp
- Industrial Hemp for renewable energy
- Ramie Fibre - Another Natural fibre:

WELLNESS FINISH WITH VITAMIN E

Functional aspects for clothing textiles, which promote the wearing comfort and well-being have become a striking sales argument. The following article introduces a product idea for "Wellness Textiles". It describes a transfer system consisting of textile/cyclodextrin-vitamin E complex.

Wellness is by long more than just a buzzword for wellbeing, vitality and fitness. It has become a social phenomenon which confers the wish of 'eternal youth' to ageing persons. The term 'wellness' has become a new life philosophy and from the commercial aspect is an important pulse generator and growth motor for many trade brances.

The 'wellness' fever has also reached the field of textile finishing and blew in some 'fresh wind' to this field with innovative product ideas. The following article describes a so-called transfer system by which vitamin E is converted from the textile onto the human skin. For thc first timc the transfer of vitamin E from the textile material into a skin model could be proven in an experiment.

VITAMIN E (A-TOCOPHEROL)

Vitamin E belongs to the group of lipid-soluble vitamins and is found in nature in many vegetable oils. The chemical term for vitamin E is "a-Tocopherol".(alpha-Tocopherol) In the cosmetic industry vitamin E is used as antioxidant and active substance among others because of its moisture binding capacity in aliphatic cosmetic creams, lotions, emulsions, body and face oils for dry skin care as well as for decorative cosmetics like lipsticks. Vitamin E is also successfully applied for various skin diseases.

Very much importance is attached to vitamin E in the food industry, too. Vitamin E is an important lipid-soluble antioxidant which has many positive physiological properties besides its vitamin character. The term 'antioxidant' describes the capability of molecules neutralizing so called radicals. Thus antioxidants are often called scavengers.

Radicals are atoms or molecules which have an unpaired electron in their outer shell. Free radicals also emerge by the normal cell breathing as side products and try to snatch away an electron from other structures for means of completing their outer shell. In this way for example the cell membrane can be damaged. Antioxidants and thus also vitamin E 'deactivate' the free radicals by giving off an electron and in this way protect the cells from 'oxidative stress'.

CYCLODEXTRINS

Cyclodextrins are circular molecules produced by the enzymatic decomposition of starch and they belong to the group of oligosaccharides and consist of 6 to 8 glucose units.

The interesting point about cyclodextrins is their cylindrical structure and the resulting properties and application possibilities. The polar OH groups of the individual glucose units are on the outside of the cylinder due to their steric arrangement. The outside is hydrophilic, whereas the inside of the cylinder is non- polar and thus hydrophobic (resp. lipophilic). The cavities of the cyclodextrins (host) can take in 'guest molecules' and release them again. The chemist calls this phenomenon host-guest-chemistry. Guests are all those molecules which could fit into the cavity and are non- polar enough to interact with the lipophilic cavity surface.

By this complexation the properties of the locked in molecules change. For example an increase of the water solubility of non- polar organic compounds or decrease of sightly volatile substances are obtained through complexation with cyclodextrins, but also an increase of the stability of the locked in substances to light, oxygen and heat.

Altogether the complexation through cyclodextrins has been researched only to a small extent, although cyclodextrins have been marketed for some time now for means of odour absorption (antismell finish). Knowledge about which substances can actually be complexed, is still very fragmentary and quite some more fundamental research will be necessary in this field.

CYCLODEXTRIN-VITAMIN-E COMPLEX

Vitamin E being a lipid soluble substance is virtually predestined for complexation with cyclodextrines. Y-cyclodextrin (8 glucose units) has proven to be a particularly suitable 'active substance'.

Examinations have shown that a 2:1 complex consisting of Y-cyclodextrin and vitamin E offers many more advantages in terms of stability than a 1:1 complex. The antioxidative potential of vitamin E is based on the reactivity of the chromanoxyl radical.

The main task of the cyclodextrine is to surround the ring system of the vitamin E molecule (benzopyran-6-ol) and continue to stabilize it. In this stabilized 'packaging', vitamin E is excellently suitable for Wellness finishes of textiles worn close to the body and these products can be purchased as 'NouWell E' from CHT R Beitlich GmbH.

FIXATION AND PROOF OF CD VITAMIN E COMPLEX ON TEXTILE SURFACES

As the CD vitamin E complex does not show any substantivity, only those processes are to be considered where the product application can be controlled like for example by padding, spraying, coating or printing. For permanent fixation of the complex, reactive polyurethanes have proven to be advantageous for the following reasons:

- Free from formaldehyde
- Soft handle
- Good permanence
- Applicable on all types of fibres.

Fixation of the complex is physical and particularly on CEL and WO chemical (reaction with -OH, -NH2-groups). Fixation of the charged cyclodextrins with reactant crosslinking agents is also be possible, but then only for application on CEL fibres. The qualitative proof of vitamin E on the textile can be done through a dyeing reaction at which the reductive properties of vitamin E are taken advantage of.

- Dripping on a $FeCl_3$-solution onto the finished textile. In the presence of vitamin E the Fe^{3+}-Ion is reduced to Fe^{2+}.
- Dripping on a dipyridyl solution. Dipyridyl forms with Fe^{2+}ions a red chelate complex.

By means of this Redox reaction vitamin E can be easily and safely seen on the fabric with the restriction that this of course can only be done on white or pastel shaded fabric

5

Textile Manufacturing

Textile manufacturing is a major industry. It is based in the conversion of three types of fibre into yarn, then fabric, then textiles. These are then fabricated into clothes or other artifacts. Cotton remains the most important natural fibre, so is treated in depth. There are many variable processes available at the spinning and fabric-forming stages coupled with the complexities of the finishing and colouration processes to the production of a wide ranges of products. There remains a large industry that uses hand techniques to achieve the same results.

PROCESSING OF COTTON

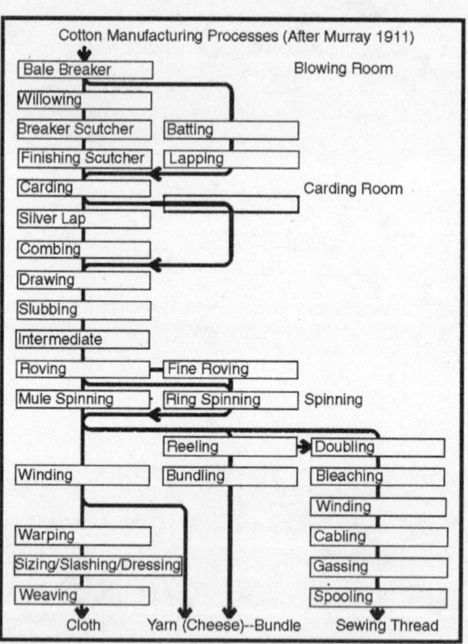

Cotton is the world's most important natural fibre. In the year 2007, the global yield was 25 million tons from 35 million hectares cultivated in more than 50 countries.

There are five stages:

- Cultivating and Harvesting
- Preparatory Processes

- Spinning
- Weaving
- Finishing

CULTIVATING AND HARVESTING

Cotton is grown anywhere with long, hot dry summers with plenty of sunshine and low humidity. Indian cotton, gossypium arboreum, is finer but the staple is only suitable for hand processing. American cotton, gossypium hirsutum, produces the longer staple needed for machine production. Planting is from September to mid November and the crop is harvested between March and May.

The cotton bolls are harvested by stripper harvesters and spindle pickers, that remove the entire boll from the plant. The cotton boll is the seed pod of the cotton plant, attached to each of the thousands of seeds are fibres about 2.5 cm long.

Ginning

The seed cotton goes in to a Cotton gin. The cotton gin separates seeds and removes the "trash" (dirt, stems and leaves) from the fibre. In a saw gin, circular saws grab the fibre and pull it through a grating that is too narrow for the seeds to pass. A roller gin is used with longer staple cotton. Here a leather roller captures the cotton.

A knife blade, set close to the roller, detaches the seeds by drawing them through teeth in circular saws and revolving brushes which clean them away. The ginned cotton fibre, known as lint, is then compressed into bales which are about 1.5 m tall and weigh almost 220 kg. Only 33per cent of the crop is usable lint.

Commercial cotton is priced by quality, and that broadly relates to the average length of the staple, and the variety of the plant. Longer staple cotton (2½ in to 1¼ in) is called Egyptian, medium staple (1¼ in to ¾ in) is called American upland and short staple (less than ¾ in) is called Indian. The cotton seed is pressed into a cooking oil. The husks and meal are processed into animal feed, and the stems into paper.

ISSUES

Cotton is farmed intensively and uses large amounts of fertilizer and 25per cent of the worlds insecticide. Native Indian variety were rainwater fed, but modern hybrids used for the mills need irrigation, which spreads pests.

The 5per cent of cotton-bearing land in India uses 55per cent of all pesticides used in India.

Before mechanisation, cotton was harvested manually and this unpleasant task was done by the lower castes, and in the United States by slaves of African origin.

DESIZING

Desizing is the process of removing the size material from the warp yarns in woven fabrics.

SIZING AGENTS

Sizing agents are selected on the basis of type of fabric, environmental friendliness, ease of removal, cost considerations, effluent treatment, etc.

NATURAL SIZING AGENTS

Natural sizing agents are based on natural substances and their derivatives:

- Starch and starch derivatives; native starch, degradation starch and chemically modified starch products
- Cellulosic derivatives; carboxymethylcellulose (CMC), methylcellulose and oxyethylcellulose
- Protein-based starches; glue, gelatin, albumen

SYNTHETIC SIZING AGENTS

- Polyacrylates,
- Modified polyesters;
- Polyvinyl alcohols (PVA),
- Styrene/maleic acid copolymers.

DESIZING PROCESSES

Desizing, irrespective of what the desizing agent is, involves impregnation of the fabric with the desizing agent, allowing the desizing agent to degrade or solubilise the size material, and finally to wash out the degradation products.

The major desizing processes are:

- Enzymatic desizing of starches on cotton fabrics
- Oxidative desizing
- Acid desizing
- Removal of water-soluble sizes

ENZYMATIC DESIZING

Enzymatic desizing is the classical desizing process of degrading starch size on cotton fabrics using enzymes. Enzymes are complex organic, soluble bio-catalysts, formed by living organisms, that catalyze chemical reaction in biological processes. Enzymes are quite specific in their action on a particular substance

A small quantity of enzyme is able to decompose a large quantity of the substance it acts upon.

Enzymes are usually named by the kind of substance degraded in the reaction it catalyzes. Amylases are the enzymes that hydrolyses and reduce the molecular weight of amylose and amylopectin molecules in starch, rendering it water soluble enough to be washed off the fabric. Effective enzymatic desizing require strict control of pH, temperature, water hardness, electrolyte addition and choice of surfactant.

OXIDATIVE DESIZING

In oxidative desizing, the risk of damage to the cellulose fibre is very high, and its use for desizing is increasingly rare.

Oxidative desizing uses potassium or sodium persulfate or sodium bromite as an oxidizing agent.

ACID DESIZING

Cold solutions of dilute sulphuric or hydrochloric acids are used to hydrolyze the starch, however, this has the disadvantage of also affecting the cellulose fibre in cotton fabrics.

REMOVAL OF WATER-SOLUBLE SIZES

Fabrics containing water soluble sizes can be desized by washing using hot water, perhaps containing wetting agents (surfactants) and a mild alkali. The water replaces the size on the outer

Scouring

Scouring, is a chemical washing process carried out on cotton fabric to remove natural wax and non-fibrous impurities (eg the remains of seed fragments) from the fibres and any added soiling or dirt. Scouring is usually carried in iron vessels called kiers.

The fabric is boiled in an alkali, which forms a soap with free fatty acids. (saponification). A kier is usually enclosed, so the solution of sodium hydroxide can be boiled under pressure, excluding oxygen which would degrade the cellulose in the fibre.

If the appropriate reagents are used, scouring will also remove size from the fabric although desizing often precedes scouring and is considered to be a separate process known as fabric preparation. Preparation and scouring are prerequisites to most of the other finishing processes. At this stage even the most naturally white cotton fibres are yellowish, and bleaching, the next process, is required.

Bleaching

Bleaching improves whiteness by removing natural colouration and remaining trace impurities from the cotton; the degree of bleaching necessary is determined by the required whiteness and absorbency. Cotton being a vegetable fibre will be bleached using an oxidizing agent, such as dilute sodium hypochlorite or dilute hydrogen peroxide.

If the fabric is to be dyed a deep shade, then lower levels of bleaching are acceptable, for example. However, for white bed sheetings and medical applications, the highest levels of whiteness and absorbency are essential.

Textile Bleaching

Textile bleaching is one of the stages in the manufacture of textiles. All raw textile materials, when they are in natural form, are known as 'greige' material(pronounced grey-sh).

This greige material will be with its natural colour, odour and impurities that are not suitable for clothing materials. Not only the natural impurities will remain on the greige material but also the add-ons that were made during its cultivation, growth and manufacture in the form of pesticides, fungicides, worm killers, sizes, lubricants, etc. The removal of these natural colouring matters and add-ons during the previous state of manufacturing is called scouring and bleaching.

BLEACHING

The next process of decolourization of greige material in to a suitable material for next processing is called bleaching. Bleaching of textiles can be classified in to oxidative bleaching and reductive bleaching.

OXIDATIVE BLEACHING

Generally oxidative bleachings are carried out using sodium hypochlorite, sodium chlorite or hydrogen peroxide. Natural fibres like cotton, ramie, jute, wool, bamboo are all generally bleached with oxidative methods.

REDUCTIVE BLEACHING

Reductive method of bleaching is done with Sodium hydrosulphite, a powerful reducing agent. Fibres like Polyamide, Polyacrylics and Polyacetates can be bleached using reductive bleaching technology.

OPTICAL WHITENERS

After scouring and bleaching, Optical Brightening Agents (OBA), are applied to make the textile material to appear more brilliant whites. These OBA are available in different tints such as blue, violet and red.

Mercerising

A further possibility is mercerizing during which the fabric is treated with caustic soda solution to cause swelling of the fibres. This results in improved lustre, strength and dye affinity.

Cotton is mercerized under tension, and all alkali must be washed out before the tension is released or shrinkage will take place. Mercerizing can take place directly on grey cloth, or after bleaching.

Many other chemical treatments may be applied to cotton fabrics to produce low flammability, crease resist and other special effects but four important non-chemical finishing treatments are:

MERCERISED COTTON

Mercerisation is a treatment for cotton fabric and thread that gives fabric a lustrous appearance. The process is applied to materials like cotton or hemp.

HISTORY

The process was devised in 1844 by John Mercer of Great Harwood, Lancashire, England, who treated cotton fibres with sodium hydroxide. The treatment caused the fibres to swell, which in Mercer's version of the process shrank the overall fabric size and made it stronger and easier to dye. The process did not become popular, however, until H. A. Lowe improved it into its modern form in 1890. By holding the cotton during treatment to prevent it from shrinking, Lowe found that the fibre gained a lustrous appearance.

PROCESS

Mercerisation alters the chemical structure of the cotton fibre. The structure of the fibre inter-converts from alpha-cellulose to a thermodynamically more favourable beta-cellulose polymorph.

Mercerising results in the swelling of the cell wall of the cotton fibre. This causes increases in the surface area and reflectance, and gives the fibre a softer feel.

THREAD

The modern production method for mercerised cotton, also known as "pearl" or "pearle" cotton, gives cotton thread (or cotton-covered thread with a polyester core) a sodium hydroxide bath that is then neutralized with an acid bath. This treatment increases lustre, strength, affinity to dye, resistance to mildew, but also increases affinity to lint. Cotton with long staple fibre lengths responds best to mercerisation. Mercerised thread is commonly used to produce fine crochet.

Singeing

Singeing is designed to burn off the surface fibres from the fabric to produce smoothness.

The fabric passes over brushes to raise the fibres, then passes over a plate heated by gas flames.

Raising

Another finishing process is raising. During raising, the fabric surface is treated with sharp teeth to lift the surface fibres, thereby imparting hairiness, softness and warmth, as in flannelette.

SINGE

A singe is a slight scorching, burn or treatment with flame. This may be due to an accident, such as scorching one's hair when lighting a gas fire, or a deliberate method of treatment or removal of hair or other fibres.

HAIRDRESSING

A singe is a treatment available at a barber's. A lit taper (candle) or other device is used to lightly burn and shrivel the hair. This is supposed to have beneficial effects - sealing cut ends, closing up the follicles, preventing the hair from bleeding and encouraging it to grow.

This practice was popular approximately a century ago; it was believed that hair had "fluid" in it and singeing would trap the fluid in. This belief has since been debunked. Singeing is still sometimes used to bond natural hair to hair extensions. Primitive cultures have also used singeing as a means to trim scalp or body hair, as a part of normal grooming or during ritual activity. Sir Francis Drake was famously said to have *singed the King of Spain's beard* when he raided Cadiz and burnt the Spanish fleet.

AGRICULTURE

In the agricultural industry, poultry and pork is singed to remove stub feathers and bristles.

TEXTILES

In the textile industry, loose fibres protruding on the surface of textile goods are singed to remove them. When done to fabrics containing cotton, this results in increased wettability, better dyeing characteristics, improved reflection, no "frosty" appearance, a smoother surface, better clarity in printing, improved visibility of the fabric structure, less pilling and decreased contamination through removal of fluff and lint.

The process is usually to pass one or both sides of a fabric over a gas flame to burn off the protruding fibres. Other methods include infra-red or heat for thermoplastic fibres. Singeing of yarns is called *"gassing"*. It is usually the first step after weaving or knitting, though the fabric may be brushed first to raise the surface fibres. Cellulose fibres such as cotton are easily singed because the protruding fibres burn to a light ash which is easily removed. Thermoplastic fibres are harder to singe because they melt and form hard residues on the fabric surface.

CALENDERING

Calendering is the third important mechanical process, in which the fabric is passed between heated rollers to generate smooth, polished or embossed effects depending on roller surface properties and relative speeds.

The calender is a series of hard pressure rollers used to form or smooth a sheet of material. In a principal application, the calender is located at the end of a papermaking process (on-line). Those that are used separate from the process (off-line) are also called *supercalenders*. The purpose of a calender is to make the paper smooth and glossy for printing and writing.

CALENDER SECTION OF A PAPER MACHINE

The calender section of a paper machine consist of a calender and other equipment the paper web is run between in order to further smooth it out, which also gives it a more uniform thickness. The pressure applied to the web by the rollers determines the finish of the paper, and there are three types of finish that the paper can have.

The first is *machine* finish, and can range from a rough antique look to a smooth high quality finish. The second is called a supercalendered finish and is a higher degree for fine-screened halftone printing. The third type of finish is called a *plater* finish, and whereas the first two types of finish are accomplished by the calender stack itself, a plater finish is obtained by placing cut sheets of paper between zinc or copper plates that are stacked together, then put under pressure and perhaps heating.

A special finish such as a *linen* finish would be achieved by placing a piece of linen between the plate and the sheet of paper, or else an embossed steel roll might be used. After calendering, the web has a moisture content of about 6per cent (depending on the furnish). It is wound onto a roll called a *tambour*, and stored for final cutting and shipping.

ETYMOLOGY

The word "calender" itself is a derivation of the word *kylindros*, the Greek word for "cylinder".

HISTORY

In the past, the paper sheets were worked on with a polished hammer or pressed between polished metal sheets in a press. With the continuously operating paper machine it became part of the process of rolling the paper (in this case also called web paper).

The pressure between the rollers, the "nip pressure", can be reduced by heating the rolls and/or moistening the paper surface. This helps to keep the bulk and the stiffness of the web paper which is beneficial for its later use.

Modern calenders have "hard" heated rollers made from chilled cast iron or steel, and "soft" rollers coated with polymeric composites. This widens the working nip and distributes the specific pressure on the paper more evenly.

SUPERCALENDER

A supercalender is a stack of calenders consisting of alternating steel and fibre-covered rolls through which paper is passed to increase its density, smoothness and gloss.

It is similar to a calender except that alternate chilled cast iron and softer rolls are used. The rolls used to supercalender uncoated paper usually consist of cast iron and highly compressed paper, while the rolls used for coated paper are usually cast iron and highly compressed cotton. The finish produced varies according to the raw material used to make the paper and the pressure exerted on it, and ranges from the highest English finish to a highly glazed surface. Supercalendered papers are sometimes used for books containing fine line blocks or halftones because they print well from type and halftones, although for the latter they are not as good as coated paper.

OTHER MATERIALS

Calenders can also be applied to materials other than paper when a smooth, flat surface is desirable, such as cotton, linens, silks, and various man-made fabrics and polymers such as vinyl and ABS polymer sheets, and to a lesser extent HDPE, polypropylene and polystyrene.

The calender is also an important processing machine in the rubber industries, especially in the manufacture of tires, where it is used for the inner layer and fabric layer.

SANFORIZATION

Sanforization is a process of treatment used for cotton fabrics mainly and most textiles made from natural or chemical fibres, patented by Sanford Lockwood Cluett (1874–1968) in 1930. It is a method of stretching, shrinking and fixing the woven cloth in both length and width, before cutting and producing to reduce the shrinkage which would otherwise occur after washing.

The cloth is continually fed into the sanforizing machine and therein moistened with either water or steam. A rotating cylinder presses a rubber band against another heated rotating cylinder, thereby the rubber band briefly gets compressed and afterwards shrinks to its final size. The cloth to be treated is transported between rubber band and heated cylinder and is forced to follow this brief expansion and recontraction and thus gets shrunk.

The bigger the pressure applied to the rubber band the bigger the shrinking afterwards.

The aim of the process is a cloth which does not shrink during clothes production by cutting, sewing or by wearing and washing the finished clothes.

For technical application cloth may be specified to have a shrink-proof value of less than 1per cent

DYEING

Dyeing is the process of imparting colours to a textile material in loose fibre, yarn, cloth or garment form by treatment with a dye. Finally, cotton is an absorbent fibre which responds readily to colouration processes. Dyeing, for instance, is commonly carried out with an anionic direct dye by completely immersing the fabric (or yarn) in an aqueous dye bath according to a prescribed procedure. For improved fastness to washing, rubbing and light, other dyes such as vats and reactive are commonly used. These require more complex chemistry during processing and are thus more expensive to apply.

DYE TYPES

For most of the thousands of years in which dyeing has been used by humans to decorate clothing, or fabrics for other uses, the primary source of dye has been nature, with the dyes being extracted from animals or plants. In the last 150 years, humans have produced artificial dyes to achieve a broader range of colours, and to render the dyes more stable to resist washing and general use. Different classes of dyes are used for different types of fibre and at different stages of the textile production process, from loose fibres through yarn and cloth to completed garments.

Acrylic fibres are dyed with basic dyes, Nylon and protein fibres such as wool and silk are dyed with acid dyes, polyester yarn is dyed with disperse dyes. Cotton is dyed with a range of dye types, including vat dyes, which are similar to the ancient natural dyes, and modern synthetic reactive and direct dyes.

HISTORY

Archaeologists have found evidence of textile dyeing dating back to the Neo-lithic period. In China, dyeing with plants, barks and insects has been traced back more than 5,000 years. Early evidence of dyeing comes from Sindh (Pakistan), where a piece of cotton dyed with a vegetable dye has been recovered from the archaeological site at Mohenjo-daro (3rd millennium BCE). The dye used in this case was madder, which, along with other dyes such as indigo, was introduced to other regions through trade. Contact with Alexander the Great, who had successfully used dyeing for military camouflage, may have further helped aid the spread of dyeing from India.

Natural insect dyes such as Tyrian purple and kermes and plant-based dyes such as woad, indigo and madder were important elements of the economies of Asia and Europe until the discovery of man-made synthetic dyes in the mid-19th century. The first synthetic dyes was William Perkins's mauveine in 1856, derived from coal tar. Alizarin, the red dye present in madder, was the first natural pigment to be duplicated synthetically, in 1869,a development which led to the collapse of the market for naturally grown madder.The development of new, strongly coloured synthetic dyes followed quickly, and by the 1870s commercial dyeing with natural dyestuffs was disappearing.

METHODS

Dyes are applied to textile goods by dyeing from dye solutions and by printing from dye pastes.

DIRECT APPLICATION

The term "direct dye application" stems from some dyestuff having to be either fermented as in the case of some natural dye or chemically reduced as in the case of synthetic vat and sulfur dyes before being applied. This renders the dye soluble so that it can be absorbed by the fibre since the insoluble dye has very little substantivity to the fibre. Direct dyes, a class of dyes largely for dyeing cotton, are water soluble and can be applied directly to the fibre from an aqueous solution. Most other classes of synthetic dye, other than vat and surface dyes, are also applied in this way.

The term may also be applied to dyeing without the use of mordants to fix the dye once it is applied. Mordants were often required to alter the hue and intensity of natural dyes and improve their colour fastness.

Chromium salts were until recently extensively used in dying wool with synthetic mordant dyes. These were used for economical high colour fastness dark shades such as black and navy. Environmental concern has now restricted their use, and they have been replaced with reactive and metal complex dyes which need no mordant.

YARN DYEING

There are many forms of yarn dyeing. Common forms are the at package form and the at hanks form. Cotton yarns are mostly dyed at package form, and acrylic or wool yarn are dyed at hank form. In the continuous filament industry, polyester or polyamide yarns are always dyed at package form, while viscose rayon yarns are partly dyed at hank form because of technology. The common dyeing process of cotton yarn with reactive dyes at package form is as follows:

- The raw yarn is wound on a spring tube to achieve a package suitable for dye penetration.
- These softened packages are loaded on a dyeing carrier's spindle one on another.
- The packages are pressed up to a desired height to achieve suitable density of packing.
- The carrier is loaded on the dyeing machine and the yarn is dyed.
- After dyeing, the packages are unloaded from the carrier into a trolly.
- Now the trolly is taken to hydro extractor where water is removed.
- The packages are hydro extracted to remove the maximum amount of water leaving the desired colour into raw yarn.
- The packages are then dried to achieve the final dyed package.¤

After this process, the dyed yarn packages are packed and delivered.

REMOVAL OF DYES

In order to remove natural or unwanted colour from material, the opposite process of bleaching or discharging is carried out.

If things go wrong in the dyeing process, the dyer may be forced to remove the dye already applied by a process that is normally known as stripping. This normally means destroying the dye with powerful reducing agents (sodium hydrosulphite) or oxidizing agents (hydrogen peroxide or sodium hypochlorite). The process often risks damaging the substrate (fibre). Where possible, it is often less risky to dye the material a darker shade, with black often being the easiest or last option.

NATURAL DYES: APPLICATION, IDENTIFICATION AND STANDARDIZATION

Natural dyes comprises of those colourants (dyes and pigments) that are obtained from animal or vegetable matter without chemical processing. They are mainly mordant dyes although some vat, solvent, pigment, and acid types are known.

Natural dyes find use in the colouration of textiles, foods, drugs, and cosmetics. Small quantities of dyes are also used in colouration of paper, leather, shoe polish, wood, cane, candles, etc.

In the earlier days, dyes were derived only from natural sources. But natural dyes suffer from certain inherent disadvantages of standardized application and the standardization of the dye itself as dyes collected from similar plants or natural sources are influenced and subjected to the vagaries of climate, soil, cultivation methods etc.

Hence for the natural dyes to be truly commercialsed and to take a competitive place with respect to the synthetic dyes, the standardization methods play a very significant and vtal role. In this paper we shall indicate some conventional and often used methods of application, the identification methods and also standardization process of the natural dyes.

Natural dyes comprises of those colourants (dyes and pigments) that are obtained from animal or vegetable matter without chemical processing. They are mainly mordant dyes although some vat, solvent, pigment, and acid types are known.

Natural dyes find use in the colouration of textiles, foods, drugs, and cosmetics. Small quantities of dyes are also used in colouration of paper, leather, shoe polish, wood, cane, candles, etc.

In the earlier days, dyes were derived only from natural sources. Some processing was required but essentially the dye itself was obtained from a plant, mineral or animal. After the accidental synthesis of mauveine by William Henry Perkin in 1856 and its subsequent commercialization, heralding the advent of coal tar dyes (now synthetic dyes), the use of natural dyes receded.

A wide range of synthetic dyestuff was thrust upon the industry and it was readily accepted for its distinct advantage over natural dyes with respect to application, colour range and availability.

Status of Natural Dyes in India: India has a very rich tradition of using natural dyes. The art and craft of producing natural dyed textile has been practiced since ages in many villages by traditional expert crafts-persons in the country. Natural dyes, when used by themselves have many limitations of fastness and brilliancy of shade.

However, when used along with metallic mordants they produce bright and fast colours. The use of metallic mordants is not always eco-friendly, but the pollution problems created by metallic mordants are of very low order and can be easily overcome.

Therefore, instead of using unsustainable technology for producing colours one can use 'Mild Chemistry' to achieve almost similar results.

In the past decade the synthetic dye manufacturing has come under the tight scrutiny of the public eye due to the following reasons:

- Certain chemical compounds used in chemical dyes are found to be carcinogenic, mutagenic and sensitizing and as a result, they have been banned. Azo dyes have been red listed. Therefore manufactures of handloom goods and those promoting their exports have come to realise that the developed countries like the US, Germany, Austria and Denmark patronizing Indian handloom prefer garments coloured with natural dyes as compared to synthetic dyes.

  This in itself provided the main thrust for the retrospection by the dyeing industry and the revival of interest in natural dyes and the introduction of eco-friendly natural dyes as an alternate to the existing synthetic dyes.
- There is a growing awareness and public concern in recent years over the issue environmental pollution and hazards arising out of the textile dyeing industry. Dyeing of fabrics with natural dyes, a rich traditional craft is now therefore being revived.

In recent years an interest in natural dyes has been manifested as,

- Reconstruction of ancient and traditional dyeing technology.
- Study of Museum textiles and textiles recovered by Archaeology.
- Conservation and restoration of old textiles.
- Growing use of natural dyes in home craft works.
- Chemical characterization of colourant in flora and fauna.
- Replacement of synthetic dyes by natural dyes for foods, safety and so forth.

Stake Holders of Natural Dyes: The use of natural dyes has increased substantially during the last couple of years.

These dyes are being mainly used by:

- Hobby groups
- Designers
- Traditional dyers and printers

- Non-government organizations (NGO's)
- Museums
- Academic institute and research associations/laboratories
- Industry.

CLASSIFICATION OF NATURAL DYES

Mainly three types of natural dye classification are done based on Chemical Class, Application Class and on Colour.

Based on Chemical Classes: The natural organic dyes and pigments cover a wide range of chemical classes. *viz.* Polymethines, Ketones, Imines, Quinines, Anthraquinonoids, Naphthoquinones, Flavones, Flavanols, Flavanones, Indigoids and Chlorophyll. Some of the important chemical classes are enumerated below.

Based on Application Class: Bancroft, in his, "Treatise on Permanent Colours" published about 160 years ago, classified natural dyes into two groups.

- *Substantive dyes*: The dyes, which dye the fibre directly, are classified as substantive dyes.
  Example: Indigo, Turmeric, Orchil.
- *Adjective dyes*: These dyes dye the material only when it is mordanted with metallic salt or with addition of metallic salt to the dye bath.
  Example: Logwood, Madder, Cochineal, Fustic.

In pure state, the adjective dyes are only slightly coloured and when used alone give poor dyeing. The above classification was replaced by an equivalent subsequent classification as Direct dyes and Mordant dyes. Another classification from Hummel is as follows

- *Monogenetic dyes*: These dyes produce only one colour irrespective of the mordant.
- *Polygenetic dyes*: The colour generated by these dyes depends on mordant used.

Another classification based on application:

- *Mordant Dyes*: Mordant dyes are defined as those dyes, which have affinity for mordanted fibres. All dyes, which form complex with mordants, are grouped under mordant dyes.
  Example: Fustic, Kermes, Cochineal etc.
  Many natural dyes are extracted along with tannin from the vegetable matter and the dye is directly absorbed by untreated cotton fibres. In such case, mordanting and dyeing takes place simultaneously. Tannin, being brownish in colour, has a tendency to modify the hue, making dull.
  In some cases, cotton fabric is treated with tannin or tannin containing vegetable matter such as harda and subsequently mordanted with metallic mordant. In this case, the tannin holds the metal and the metal forms complex with dye.

- *Vat Dyes*: The 'Vat Dye' derived their name from the wooden fermentation vessel called 'Vat', which was at one time used for reducing the dye in order to convert it into soluble form. This process of solubilization is called Vatting and the soluble form of dye is called leuco dye. The soluble leuco form has affinity for natural fibres, which can be oxidized back on fibre on exposure to air. The vatting is also carried out by treatment with reducing agent, such as sodium hydrosulphite and alkali, such as sodium hydroxide.
  Example: Indigo
- *Direct Dyes*: *e.g.* Turmeric, which though a fugitive dye has an affinity for cotton. This dye is also directly sorbed by other natural fibres.
  Examples: Turmeric, Annatto, Harda, pomegranate (Punica granatum L.), Carthamin obtained from Safflower etc.
- *Acid Dyes*: Acid dyes are type of direct dyes for polyamide fibres like wool and silk. These dyes are applied from acidic medium and they have either sulphonic or carboxylic group(s) in the molecule.
  Example: Saffron.
  An aftertreatment with tannic acid and tartar emetic, known as back tanning, improves the wash fastness of these types of dyes.
- *Basic Dyes*: These are cationic dyes, which on ionization gives coloured cations. They are used on polyamide fibres such as wool and silk from neutral or mildly acidic conditions. These dyes may be applied on cotton mordanted with tannic acid and tartar emetic or other metal salts. These dyes have low light fastness.
  Example: Barberine.
- *Disperse Dyes*: A disperse dye has a low relative molecular mass (r.m.m.), low solubility and no strong solubilising groups. These dyes have hydroxyl and/or amino groups which imparts some solubility to the dye molecule. One such dye could be Lawsone. Many other flavone and Anthraquinone dyes can qualify to be classified as Disperse dyes.

C.I.	Natural No. of Dyes	Per cent
Yellow	28	30.4
Orange	6	6.5
Red	32	34.8
Blue	3	3.3
Green	5	5.5
Brown	12	13.0
Black	6	6.5

Based on Colour: In the Colour Index, the dyes are classified according to chemical constitution as well as major application classes. Within application class, the dyes are arranged according to hue. Natural dyes form a separate section. The no of dyes in each hue are given below. Some dyes produce more than one hue.

Of the 92 natural dyes listed in Colour Index, chemical structure of 67 dyes is disclosed. Many dyes have more than one compound and some dyes have identical structures.4

SOME IMPORTANT NATURAL DYES

Colouring matter could be obtained from all almost all vegetable matter. However, only a few of these sources yield colourants which can be extracted and work out to be commercially viable. Similar is the case of colourants obtained from animal origin.

Basically, three primary colours are required to get any given hue (or colour). This type of approach has been worked out for synthetic dyes. However, in the case of natural dyes, the dyeing procedures are different for different dyes and they cannot be blended to get the required colour easily. Never-the-less, while looking for different colours, it is better to have a limited number of dyes with good fastness properties rather than having too many colours (sources) with limited fastness properties. While selecting the proper palette of colours, one would like to have atleast one blue, one red and a yellow to start with. Due to the limited number of natural dyes available, the correct choice of the dyes is very important. Information on some important natural dyes is given below.

- *Blue Dyes*: The Colour Index lists only three natural blue dyes; namely Natural Indigo, sulphonated natural Indigo and the flowers of the Japanese 'Tsuyukusa' used mainly for making awobana paper. The only viable choice among the blue natural dyes is Indigo. Natural indigo is obtained by fermenting the leaves of various species of Indigofera, running off the liquor and oxidizing it to precipitate the dye. Woad (Isatis Tinctoria L.) is another source of indigo. The main ingredients of natural indigo are Indigotin and Indirubin. Natural indigo has higher affinity and the dyed fabrics have better fastness.
- *Red Dyes*: The Colour Index lists 32 red natural dyes. The prominent among them are Madder. All these dyes are based on anthraquinone molecule except Brazil and Sappan wood based dyes. These dyes are prone to oxidation and hence are not suitable.

  The strong and almost fadeless cotton dye known as Turkey Red was developed in India and spread from there to Turkey. It involved about twenty separate processes using wood, oil and rancid fat, charcoal, cow/sheep/dog dung, and liquid content of animal stomach. Only the dyers and their families not surprisingly, occupied villages where the process was carried out. The use of madder declined abruptly following the development of synthetic alizarin in 1869.

- *Yellow Dyes*: Yellow is the most common colour in the natural dyes. However, most of the yellow colourants are fugitive. The Colour Index lists 28 yellow dyes. Some of the important yellow dyes are obtained from Barberry (*Berberis aristata*), Tesu flowers (*Butea frondosa, Monosperma*) and Kamala (*Mallotus philippensis*).

  Other sources of yellow dyes are Black oak (*Quercus velutina*), Turmeric (*Curcuma longa*), Weld (*Reseda luteola*) and Himalayan rhubarb (*Rheum emodi*).

Mordant: Mordant is a chemical, which can fix itself on the fibre and combines with the dyestuff. A link is therefore formed between the dyestuff and the fibre, which allows certain dyes with no affinity to be fixed on to the fibre.

Types of Mordants: Three types of mordants namely, Metal Salts or Metallic Mordants. Several different metal salts can be used for mordanting.

The five most effective ones are,

1. *Alum*: Potassium aluminum sulphate
2. *Copper*: Copper sulphate
3. *Chrome*: Potassium dichromate
4. *Iron*: Ferrous sulphate
5. *Tin*: Stannous chloride, Stannic chloride.

TANNINS AND TANNIC ACID

Tannins are naturally occurring compounds high mol. wt (500-3000) containing phenolic hydroxyl groups (1-2 per 100 Mw) to enable them to form effective crosslinks between proteins and other macromolecules. The vegetable tannins divided structurally into two distinct classes depending on the type of phenolic nuclei involved and the way they are joined together.

- Hydrolysable tannins obtained from myrobalan fruit, oak bark, gallnuts, Pomegranate rind, sumac leaves.
- Condensed tannins like catechin obtained from acacia catechu.

  Tannins such as harda, tannic acid, etc are considered as natural mordants.

  Gallnut contains 60-77per cent of Tannic Acid.

OIL MORDANTS

Oil –mordants are mainly used in the dyeing of Turkey Red Colour from madder. The main function of the oil-mordant is to form a complex with alum used as the main mordant. Since alum is soluble in water and does not have affinity for cotton, it is easily washed out from the treated fabric. The naturally occurring oils contain fatty acids such as palmitic, stearic, oleic, ricinolic etc and their glycerides.

The –COOH groups of fatty acids react with metal salts and get converted into –COOM, where M denotes the metal, for instance in the case of alum it would be Al. subsequently, it was found that the treatment of oils with concentrated sulphuric acid produces sulphonated oils which possess better metal binding capacity than the natural oils due to the

introduction of sulphonic acid group, -SO3H. The sulphonic acid can react with metal salts to produce –SO3M. The bound metal can then form a complex with the mordant dye such as madder to give Turkey Red colour of superior fastness and hue.

Techniques of Mordanting: A subsequent treatment of the tannin treated substrate with metal salts such as alum introduces aluminum ions in the fibre. The tannin treated material at the hydroxyl or carboxyl groups sorbs these ions, either by forming a metal-complex or metal salt.

These metal ions then provide sites for mordant dyes. Hence introducing metal-ions in the fibre. Either directly or as tannin-metal complexes can increase the affinity of the substrates towards natural dyes. Therefore, tannins by themselves do not act as mordants but the tannin-metal salt combination acts as a mordant for the natural mordant dyes.

Methods of Mordanting: The three methods used for mordanting are,

- *Pre-mordanting*: The substrate is treated with the mordant and then dyed.
- *Meta-mordanting*: The mordant is added in the dye bath itself.
- *Post-mordanting*: The dyed material is treated with a mordant.

The methods have different effects on the shade obtained after dyeing and also on the fastness properties. It also depends upon the dye and the substrate. It is therefore necessary to choose a proper method to get the required shade and fastness by optimization of parameters. 3, 4.

NATURAL DYES-"TECHNIQUES OF EXTRACTIONS"

Natural Dyes cannot be used directly from their renewable sources. Using raw materials for dyeing has many limitations. Safe and cheap extraction of main colouring component is most important without affecting the extraction conditions and avoiding any contamination in various extraction techniques. Several Extraction Methodologies for natural dye that comply with both consumer preference and regulatory control and that are cost effective are becoming more popular.

Techniques of Extractions of Natural Dyes involve,

- 'Simple Aqueous Methods'
- 'Complicated Solvent Systems'
- 'Ultrasonic Extraction in Sonicator'
- 'Supercritical Fluid Extraction Techniques'.

Simple Aqueous Methods: Natural dyes are mixed with required amount of water and boiling for optimum time (found out by optimization of parameters) which is 60 minutes in most cases. The content is cooled to Room Temperature and filtered. The filtrate is used as a dye for dyeing.

Example: Extraction of Onion Dye: The outermost dry papery skins of onion were removed and boiled with water for 1 hour. The content was cooled to a R.T. and filtered. The filtrate was used as the dye solution.

COMPLICATED SOLVENT SYSTEMS

The dried material (leaves, roots, barks, wood, resinous secretion of insects etc) are ground to very fine particles. The crude dried powder is weighed and solvent extracted using Soxhlet Apparatus,Steam Heated Extractor. Different solvents (such as Acetone, chloroform, ether, n-hexane, alcohol, soda ash, etc.) are used for Extraction. The process is carried out for 4 hours. The dye extract is evapourated in an evapourating dish over a water bath. After evapourating to dryness, the solute is weighed and the percentage yield is calculated.

Example: - Extraction of Henna (lele) using solvent: - The dried leaves were ground to very fine particles. 40 gms of the crude dried powder was weighed and solvent extracted using a Soxhlet apparatus. Three different solvents acetone, chloroform and water were used for the extraction. 250 ml of each solvent was used. The process was carried out for four hrs. The dye extract was evapourated in an evapourating dish over a water bath. After evapourating to dryness, the solute was weighed and the percentage yield is calculated.10

ULTRASONIC EXTRACTION IN SONICATOR

The combined effect associated with Ultra-Sound Energy is of cavitations, compressions, rarefactions, and microstreaming results in intermolecular tearing and surface scrubbing. In particular, it has been noted that, some reactions when exposed to ultrasonic energy become faster with lower temperature that is the most beneficial effect as it reduces processing time and energy consumption and improves product quality in the colouration of textiles.

Example: Sappanwood extract is obtained from boiling the matured wood in hot water from the chips or rasped wood and extraction is preferably done on counter current principle, but even boiling under pressure gives stronger decoction. However, this procedure has serious defects heat changes the colour of red dye to brown.

To overcome this drawback we use Sonicator for extraction of dye in water using ultra-sound energy. Optical Density is recorded to evaluate the relative dye content. The extraction of the dye by usual technique required long hours of immersion of the wood chipping in water or heating. Both these processes are causing the dye to decompose due to oxidation. To extract the dye faster and in a more efficient manner Sonicator with Ultra-Sound Energy of 20 kHz frequency is used. Due to its fugitive character, Sappanwood has found little use, only some cottage dyers in some remote villages found use of this where synthetic dyestuff was not available. Ultrasonic Extraction Followed by

Microwave Extraction for solids found extensive use but mainly based on organic solvent extraction.

SUPERCRITICAL FLUID EXTRACTION TECHNIQUES

No doubt, safety of both producers and consumers is now a major requirement of any new product or process. Accordingly, compelling regulations on the usage of hazardous, carcinogenic, or toxic solvents as well as high energy costs for solvent regeneration have curtailed the growth of the natural extract industries.

To suppress the competitive edge of synthetic materials, alternative extraction methodologies are used. One of such major technologies that have emerged over the last two decades as the alternative to the traditional solvent extraction of natural products is the Supercritical Fluid Extraction Techniques.

It uses a clean, safe, inexpensive, non-flammable, non-toxic, environment-friendly, non-polluting solvent, such as CO_2. Besides the energy cost associated with this Novel Extraction Technique are lower than the costs for traditional solvent extraction method. Supercritical Fluid Extraction Technology is thus increasingly gaining importance over the conventional techniques for extraction of natural products.

A large body of experimental data has been accumulated on the solubility and extractability of natural products, such as steroids, alkaloids, anticancer agent, oils from seeds, and caffeine from coffee beans, in various SCF solvent such as CO2, ethane, ethylene, and N2O. CO_2 is the most widely investigated SCF solvent since its Critical Temperature ($Tc = 31.1°C$) makes it an ideal solvent for extracting materials that are thermally liable. Supercritical fluid extraction is an advanced separation technique based on the enhanced solvating power of gases above their critical point. CO_2 is an ideal solvent in the food, dye, pharmaceutical and cosmetic industries, where it is essential to obtain final products of a high degree of purity.

Example: Supercritical CO_2 Extraction of lycopene and β-carotene from ripe tomatoes. Main pigments of ripe tomatoes are the carotenes, compounds of colours ranging between yellow and red, *i.e.* α, β and γ-carotene, lycopene and xanthophylls at very low concentrations.

Experimental: Supercritical CO_2 extraction are performed by an SFE-400 thermal pump (Supelco), equipped with a 10 cm3 internal volume extractor. The extraction pressure could be adjusted in 100 psi increment upto a maximum pressure of 4000 psi.

The temperature range is 30- 200 °C. The extracted substances are recovered in a vial connected to a restrictor. At each run, the rate flow is maintained at about 500 cm3/min. Several methods are used for the total or partial removal of water from the tomato samples submitted to the supercritical fluid.

In all cases, thesamples are protected from the action of both light and oxygen in the air in order to prevent them from damaging the dye. The extractions are performed on ripe tomato pulp or dry skins, since they contain about 5 times as much lycopene as in the whole tomato pulp. The extractions are initially conducted on completely fresh tomatoes, ground and filtered in a vacuum. In order to adsorb the remaining moisture, the samples are mixed with silica gel.

The measurements are carried out on 'camone' greenhouse tomatoes, grown in Sardinia and on field tomatoes. Extractions are subsequently carried out on the skins and dried seeds of field tomatoes. The tomato skins and seeds are dried for 24 hours in an air drier at 35 °C and stored at -5 °C, the product is ground before being extracted.

The extractions are run by submitting 2.5gm of fresh or dry ground sample to different pressures (2500-4000 psi) and temperatures (40-80 °C) for 30 min. both in the presence (1ml) and absence of an entrainer. The extract coming out of the restrictor are collected in ethanol and analysed by HPLC.

IDENTIFICATION OF NATURAL DYES

Today natural dyes are used in limited quantities by craftsman in various parts of the world. Although difficult to obtain commercially, dyes are readily obtained from plant sources *viz.*, flowers, leaves, barks, seeds, roots, etc. Craftsman are becoming gradually enthusiastic about this out-dated and time consuming process for one of the reasons that natural dyestuffs procedure offbeat, one-of-a kind colour.

There are not two lots identical, each having subtle differences due to impurities peculiar to the particular plant material used. Thus, the very characteristics of natural dyes that made the early dyers despair, appeal to today's craftsman searching for the unique.

Since last decade application of natural dyes on cotton, silk, wool, jute is gaining popularity all over the world possibly due to the context of German ban on synthetic Azo dyes which are based on carcinogenic or allergic arylamines. Natural dyes derived from flora and fauna are believed to be safe but sometimes the method of application using metallic mordants makes them non-ecofriendly to humans and environment.

However, natural dyes by virtue of their unique colour/shade and aesthetic characteristics have a potential export market. Recently, the buyers from EEC countries especially from Germany insist on a certificate for eco-friendliness of the commodities and therefore, testing of natural dyed substrates as per Eco-standard as well as their anthenticity has become imperative.

Literature survey reveals that some work has been reported an extraction and identification of natural dyes from historic and archeological textiles but a systematic approach to extract and identify the range of colours in natural dyes from dyed textiles, is not readily available.

- Thin Layer Chromatography (TLC) was used by many workers to identify various natural dyes in textiles. The dyes detected were based on insect and vegetable yellow, red and blue colour.
- High Performance Liquid Chromatography (HPLC), which is a very useful tool to identify synthetic as well as natural dyes, was also used.
- Infra-Red Spectral studies were carried out to identify natural dyes *viz.*, alizarin, indigo, dibromoindigo, saffron, weld, Persian berries and safflower yellow.
- While a non-destructive method was devised, for identifying faded dyes on fabrics through examination of their Emission and Excitation Spectra.
- UV/Visible Spectroscopic studies to identify a number of natural dyes were carried out by Schweppe, sobko and Chikh were carried out to identify madder, cochineal, indigo, etc., using different solvents for extraction.

TECHNIQUES OF IDENTIFICATION

By Thin Layer Chromatography (TLC): Thin layer chromatography is a microanalytical technique, which separates by adsorption, partition or ion exchange on a thin layer or film of adsorbent coated on a glass plate. This technique was introduced in 1938 by two Russian workers Ismailov and Shraiber, but only since 1958 has it rapidly gained popularity as a result of the work of Stahl, a German pharmacy lecturer, who developed, among other things, a spreading device to produce uniform layers of adsorbent on glass plates. He also standardized the preparation of suitable adsorbents to give consistent results. One of the typical terms used in TLC is,

- *Column Capacity Ratio (K')*: K' value of solute is the usual method of indicating solute retention.

  K' = (Tr-To)/To

  Where,

  Tr = retention time of given peak

  To = retention time of unretained (solvent) peak

  Use of K' values is preferred to simply quoting retention times, since latter can vary with flow rate variation from day to day, while K' remains constant.

  K' values in HPLC are related to Rf values in TLC.

  K' = (1- Rf)/Rf

  Rf = 1/(1 + K')

  Rf = Ratio of distance traveled by solute to distance traveled by solvent front. K' value measures ratio of time spent by solute in stationary phase to time spent in mobile phase.
- The band of aqueous solution of the dye is applied on TLC plates by using TLC sampler. Development of the band is done in an ascending

developing chamber with a mixture of different solvent in a required proportion as a solvent system. Plates dried using a drier and the bands are visualized in UV lamps and then scanned in a TLC scanner.

- TLC studies are carried out on the extract of dyes using natural dye standards as reference.

High Performance Liquid Chromatography (HPLC): It can separate a wide variety of compounds ranging in molecular weight from less than a hundred to upwards of few million. It is particularly applicable to the fractionation of highmolecular weight substances, thermally unstable compounds and products, which cannot be volatilized without decomposition.

High performance liquid chromatography (HPLC) is a powerful analytical tool and involves a hydrophobic, low polarity stationary phase that is chemically bonded to an inert solid such as silica. The process is conducted at a high velocity and pressure drop. The column is shorter and has a small diameter, but it is equivalent to possessing a large number of equilibrium stages.

The separation is essentially an extraction operation and is useful for separating non-volatile components. If two dyes have same Rf value, they are difficult to analyse by TLC.

In contrast, dyes of different colours, which have the same retention time, may be quantitated with HPLC provided their spectra do not interfere. For example, many green dyes are mixtures of blue and yellow dyes and both dyes may be quantitated by HPLC even when they have the same retention time.

Example: lycopene and β-carotene are determined on a Hewlett Packard 1050 HPLC, equipped with an autoinjector and UV-Visible Detector, Spherisorb column C18, 5 mm, 25 cm × 4.6 mm. a mixture of methanol, THF and water in a 67:27:6 ratio with a flow of 1ml/min (2 μl injection volume) is used as the mobile phase; the detection is performed at 446 nm, corresponding to maximum of lycopene.

Lycopene and â-carotene were identified by comparing the retention time of the two pigments in the extraction mixture with those of their respective standard compounds. (Sigma products.)

Lycopene was also identified by comparing the UV-Visible, 'H NMR and Mass Spectra of the extracted substance with those of standard lycopene. The amount of total extractable pigment was determined after extraction with acetone: hexane for six hrs using a Soxhlet apparatus.12,17.

Infra-red Spectrosopy: IR spectroscopy is routinely used for identification of samples in the gas, liquid and solid states. Sample cells are made from materials, such as NaCl and KBr, that are transparent to IR Radiation.

In comparison to other IR instruments, Fourier Transform Infra-Red Spectrometers provides for rapid data acquisition, in signal-to-noise ratio through signal averaging. In FTIR, the monochromator of IR instruments is replaced with a interferometer.

Example: About 2mg of the purified dye sample is weighed in 298mg KBr pellets and the FTIR spectra are recorded between 4000 to 400 cm-1 range on a Perkin Elmer system 2000FTIR.

UV-Visible Spectroscopy: Absorption spectroscopy provides a useful tool for Qualitative Analysis, Identification of a pure compound by this method involves an empirical comparison of the details of the spectrum (maxima, minima and inflection points) of the unknown with those of pure compounds; a close Matc h is considered to be good evidence of chemical identity, particularly if the spectrum of unknown contains a number of sharp and well-defined peaks. Dyes are extracted from the dyed textiles using DMF as solvent. The UV-Visible spectra of the extract are recorded with natural dye solution of DMF as reference standard. UV-Visible Spectrophotometer model: - Hitachi U-2000 and GBC UV/VIS 918..

Differential Scanning Calorimetric Studies: Differential Scanning calorimetric (DSC) studies are carried out to investigate the thermal behaviour of the natural dyes. Dye powder is directly used for DSC Studies. Identification of different natural dyes is carried out by comparing the details of endothermic and exothermic peaks from DSC Thermogram, and values of ÄH for various natural dyes. DSC model: Mettler DSC 30.

Analysis of Red-listed Chemicals: The extracted dye is tested for the presence of Toxic Metals, Pesticides and Banned Aryl Amines by the methods prescribed for synthetic dyes.

Heavy Metals: Wet Ashing of Extract and heavy metal content is determined by Atomic Absorption Spectroscopy (AAS).

Pesticides: Extraction, clean up and Pesticides content is determined by Gas Chromatography/Electron Capture Detector (GC/ECD), High Pressure Thin Layer Chromatography (HPTLC).

Banned Aryl Amines: Reductive, cleavage, isolation and it is determined by Gas Chromatography/Mass Spectroscopy (GC/MS.)

Mordants: According to eco-standards, copper and chrome are red- listed and hence are not to be used. Problem posed by a mordant is that a substantial proportion of it is left unexhausted in the residual dyebath and may cause serious effluent problems.

Aluminum and iron are relatively innocuous; they are abundantly available and produce excellent dyeings.

STANDARDIZATION OF NATURAL DYES

Dyes of natural origin are great for colour appreciation as any variation in the concentration of dye, mordant, types of water, soil and climate give variation in colours.

In recent years, natural dyes are again getting importance due to harmful effects caused by synthetic dyes during their production and use.

TECHNIQUES OF STANDARDIZATION

Standardization of natural dye should be done by conversion of natural dye extracts into a fine powdered, paste or solution form. Only water as such or at different pH and temperature would be used for the colouring matter from the raw materials. Parameters such as pH, time and temperature of extraction should be worked out for optimization to achieve standardization.

Standardized natural dye is should be pure and does not change appreciably due to atmospheric conditions. For the determination of standardized natural dyes accurately, it is necessary to compare it with some known standard of purity Standardization includes determination of extractive values using various solvent.

- *Ash values*: Determination of ash value is useful for determining low-grade extracts.
- Rf values: - Paper chromatography and Thin Layer Chromatography is widely used as a mean of assessing Quality and Purity. Rf values of natural dyes are determined under specific conditions and can be used for standardization by an aid to identification.
- Most of natural dyes are pH sensitive and they change colour with changing pH, use of buffered solution for extraction of the dye is first step towards standardization.
- Same natural colour extracts are not give the original colour of extract from the colour obtained on the fabric. Therefore, reproducibility of natural dye shade is very important in the process of standardization.
- Parameters such as pH, Temperature of extraction and conditions for getting the maximum colouring matters should be standardized.
- Considerable Standardization with respect to moisture, dye and ash content, particle size, the pH of dye storage stability should be done.

EXACT MEASUREMENT IS THE KEY TO STANDARDIZATION

Standardization of Dye Extraction: The natural dye source (leaves, roots, woods etc) was collected, dried in shade and powdered sample is boiled in various solution medium (aqueous, alkaline, acid etc) in order to optimize the method of extraction.

The optimizations are based on the optical density of the liquid, before and after dyeing using spectrophotometer and visual appearance.

Optimization of Alkali/Acid concentration: Optimum alkali/acid concentration for 100 ml of water, for a pH to be standardized.

Optimization of Extraction Time and Dyeing Time: Boiling for longer duration increases dye concentration. High dye yield is found after boiling for optimized time. High absorption is found for optimized dyeing time, above which dye depletion is observed.

Optimization of concentration of Mordants: The concentration of mordants is optimized by considering the visual appearance and dye absorption.

Optimization of Mordanting Methods: Sometimes more than the mordant, the mordanting methods contributes towards the production of darker or lighter and faster shades.

Project on Standardization of Eco-Friendly Natural Dyes: Alps Industries Limited. was conceived and submitted a project for standardization of ecofriendly natural dyes to Technology Information, Forecasting and Assessment Council (TIFAC) a wing of Department of Science and Technology GOI (DST.) The total value of project was ₹ 4.25 Crores, out of which TIFAC was to give a lone of ₹ 2 Crores over a period of 3 years and the remaining contribution was to come from the Alps Industries Ltd.

Project finally approved by the TIFAC under its Home Grown Technology scheme in July 1998. A highly competent team of Technologists from Academic Institutes and Industry are monitored the project.

A range of Do-It-Your-Self natural dye kits have been created by Alps Industries Ltd.

Kit not only have the purified natural dye powder and required Auxiliaries and mordants but also have an instruction book with dyeing material and simple glasswares21, 22..

Example: Standardizing dyeing condition for African Marigold.

Method of Dye Extraction: Known quantity of dye Marigold petals (African Marigold variety) were taken and soaked in cold water (MLR 1:40) overnight. The petals were boiled in the same water.

The Optimum Time for the extraction of dye was determined by boiling the dye material for 30, 45, 60 and 75 min.

The pH was recorded after dye extraction and then the pH was changed to acidic (pH 4) or alkaline (pH 10) by adding Acetic acid or Potassium hydroxide respectively, depending upon the medium which dyeing was to be carried out. Finally the dye extract solution was filtered and Optical Density was measured.

Dye uptake: Relation of wavelength (γ). The dye extract was subjected to visual light of wavelength (γ) using ELICO-SL 171 minispectrophotometer. The wavelength at which the optical density obtained was taken as the suitable wavelength, for the colour of the dye.

OPTICAL DENSITY BEFORE DYEING

Percentage of dye absorption is calculated by Optical density _ Optical Density before Dyeing after Dyeingper cent of Dye Absorption = * 100.

Dilution Factor: Dilution at which O.D. was around 0.3 was selected as a suitable dilution. Therefore, Dilution Factor for Marigold is 10.23.

Thus, the Standardization leads to perfection in natural dyed product with respect to overall Fastness Properties.

CONCLUSION

- In recent years, an interest in natural dyes has been mainly manifested as Conservation and Restoration of old Textiles with replacement of synthetic dyes (which uses violent technology) by natural dyes for textiles, food, safety and so forth by using Mild Chemistry.
- The buyers from European Economic Community (EEC) Countries especially from Germany insist on a certificate for eco-friendliness of the commodities and therefore, Identification and Testing of natural dyes and natural dyed substrates as per Eco-standards as well as their anthenticity has become imperative.
- The Research and Development work in Standardization of natural dyes are least or negligible. Very few serious attempts have been made to generate new information on the use of natural dyes. Most of the researches in this area are carried away by the empirical information reported in the literature that does not have any scientific reasoning or basis. Using raw materials for dyeing has many limitations. Besides its being of unknown composition, it generates considerable amount of biomass that is cumbersome to handle in the dye-house. Non-availability of natural dyes, especially in the standardized form, which may be powder, paste or solution form.
- If there had been significant research on the use of natural dyes, it would be probable that they would be already being much more widely used than they currently are. As there is much catching up to do after 150 years of neglect, there is rapid scope for developments. This applies to the techniques of agricultural production and processing as well as to dyeing itself. A medium-to-long-term view of research and development is needed. It is unlikely that the total cost of R and D required to make natural dyes a profitable commercial reality need be very great when compared with many industrial products.
- Only small portion of a huge icebergs of standardization of natural dyes are only achieved. The contributions for standardization of natural dyes are made by some companies like Alps Industries Ltd. that uses Supercritical CO_2 plant for extraction of dyes for a step towards standardization. The standardized natural extracts are very much useful for Textiles, food, pharmaceuticals and cosmetics.
- It is a challenge for any enlightened dyer for working out suitable standardized applications of natural dyes on Textiles with eco-friendliness.

BRITISH STANDARD YARN MEASURES

- 1 thread = 55 inches (about 137 cm)
- 1 skein or rap = 80 threads (120 yards or about 109 m)

- 1 hank = 7 skeins (840 yards or about 768 m)
- 1 spindle = 18 hanks (15,120 yards or about 13.826 km)

JUTE

Jute is a bast fibre, which comes from the inner bark of the plants of the Corchorus genus. It is retted like flax, sundried and baled. When spinning a small amount of oil must be added to the fibre. It can be bleached and dyed. It was used for sacks and bags but is now used for the backing for carpets. Now a recent days Jute-Cotton blended fabrics production processes is innovated by Bangladeshi scientists where Jute fibre are blended with cotton and fabricated for the production of various diversified products such as Home Textiles and different decorative fabrics. Dr.Siddique Ullah, Dr. Monjur-e-Khuda and others from BCSIR and BJRI (Bangladeshi Jute Research Institute) improve the chemical treatment method. In addition high value added shutting shirting are made from this fabric with definite woolenising, bleaching, dyeing, printing and finishing. This Jute-Cotton blended fabrics, in the 1970s, were known as Jutton fabrics.

HEMP

Hemp is a bast fibre from the inner bark of Cannabis sativa. It is difficult to bleach, it is used for making cord and rope.

- Retting
- Separating
- Pounding

OTHER BAST FIBRES

These bast fibres can also be used: kenaf, urena, ramie, nettle.

OTHER LEAF FIBRES

Sisal is the main leaf fibre used; others are: abacá and henequen.

PROCESSING OF PROTEIN FIBRES

Wool

Wool comes from domesticated sheep. It forms two products, woolens and worsteds. The sheep has two sorts of wool and it in the inner coat that is used. This can be mixed with wool that has been recovered from rags. Shoddy is the term for recovered wool that is not matted, while mungo comes from felted wool. Extract is recovered chemically from mixed cotton/wool fabrics.

The fleece is cut in one piece from the sheep.This is then skirted to remove the soiled wool, and baled. It is graded into long wool where the fibres can be up to 15 in, but anything over 2.5 inches is suitable for combing into worsteds.

Fibres less than that form short wool and are described as clothing or carding wool.

At the mill the wool is scoured in a detergent to remove grease (the yolk) and impurities. This is done mechanically in the opening machine. Vegetable matter can be removed chemically using sulfuric acid (carbonising). Washing uses a solution of soap and sodium carbonate. The wool is oiled before carding or combing.

- *Woollens*: Use noils from the worsted combs, mungo and shoddy and new short wool
- *Worsteds Combing*: Oiled slivers are wound into laps, and placed in the circular comber. The worsted yarn gathers together to form a top. The shorter fibres or noils remain behind and are removed with a knife.
- Angora

Silk

The processes in silk production are similar to those of cotton but take account that reeled silk is a continuous fibre. The terms used are different.

- *Opening bales, Assorting skeins*: where silk is sorted by colour, size and quality, scouring: where the silk is washed in water of 40 degrees for 12 hours to remove the natural gum, drying: either by steam heating or centrifuge, softening: by rubbing to remove any remaining hard spots.
- Silk throwing (winding). The skeins are placed on a reel in a frame with many others. The silk is wound onto spools or bobbins.
  - Doubling and twisting. The silk is far too fine to be woven, so now it is doubled and twisted to make the warp, known as organzine and the weft, known as tram. In organzine each single is given a few twists per inch (tpi), and combine with several other singles counter twisted hard at 10 to 14 tpi. In tram the two singles are doubled with each other with a light twist, 3 to 6 tpi. Sewing thread is two tram threads, hard twisted, and machine-twist is made of three hard-twisted tram threads. Tram for the crepe process is twisted at up to 80 tpi to make it 'kick up'.
  - Stretching. The thread is tested for consistent size. Any uneven thickness is stretched out. The resulting thread is re-eled into containing 500 yd to 2500 yd. The skeins are about 50 in in loop length.
  - Dyeing: the skeins are scoured again, and discolouration removed with a sulphur process. This weakens the silk. The skeins are now tinted or dyed. They are dried and rewound onto bobbins,

spools and skeins. Looming, and the weaving process on power looms is the same as with cotton.

- Weaving. The organzine is now warped. This is a similar process to in cotton. Firstly, thirty threads or so are wound onto a warping re-el, and then using the warping re-els, the threads are beamed. A thick layer of paper is laid between each layer on the beam to stop entangling.

DISCUSSION OF TYPES OF MAN MADE FIBRES

Synthetic fibres are the result of extensive development by scientists to improve upon the naturally occurring animal and plant fibres. In general, synthetic fibres are created by forcing, or extruding, fibre forming materials through holes (called spinnerets) into the air, thus forming a thread. Before synthetic fibres were developed, cellulose fibres were made from natural cellulose, which comes from plants.

The first artificial fibre, known as art silk from 1799 onwards, became known as viscose around 1894, and finally rayon in 1924. A similar product known as cellulose acetate was discovered in 1865. Rayon and acetate are both artificial fibres, but not truly synthetic, being made from wood. Although these artificial fibres were discovered in the mid-nineteenth century, successful modern manufacture began much later in the 1930s. Nylon, the first synthetic fibre, made its debut in the United States as a replacement for silk, and was used for parachutes and other military uses.

The techniques used to process these fibres in yarn are essentially the same as with natural fibres, modifications have to be made as these fibres are of great length, and have no texture such as the scales in cotton and wool that aid meshing.

Bibliography

Ainley, Albert: *Woolen and worsted loomfixing; a book for loomfixers and all who are interested in the production of plain and fancy worsteds and woolens*, Lawrence Mass, 1900.

Alford, Marianne Margaret Compton Cust: *Needlework as art*, London: Sampson Low Marston Searle and Rivington, 1886.

Alfred Jenks & Son: *Illustrated Catalogue of Machines Built by Alfred Jenks & Son, Manufacturers of Every Variety of Cotton and Wool Carding, Spinning, and Weaving Machinery, in All Its Departments*, Bridesburg, Pa.: Alfred Jenks & Son, 1853.

Allen, Frederick James: *The shoe industry*, Boston: The Vocation bureau of Boston, 1916.

Allen, Nellie B: *Cotton and other useful fibers*, Boston: Ginn, 1929.

Allington, Sara May: *Practical sewing and dressmaking*, Boston: Page Co., 1913.

Allinson, May: *Dressmaking as a trade for women in Massachusetts*, Washington: Govt. print. off., 1916.

Atkinson, Edward: *Cotton and Cotton Manufactures in the United States*, Boston: Franklin Press, 1880.

Baird, Robert H: *The American Cotton Spinner, and Managers' and Carders' Guide*: A Practical Treatise on Cotton Spinning, Philadelphia: H. C. Baird, 1869.

Batchelder, Samuel: *Introduction and Early Progress of the Cotton Manufacture in the United States*, Boston: Little, Brown and Company, 1863.

Brown Brothers & Co: *Brown Brothers & Co., Successors to Butler, Brown & Co.: Manufacturers of Shaw's U. S. Standard Ring Travelers, Belt Hooks, Wire Goods, Etc., and Dealers in Supplies for Cotton, Woolen, Silk, Jute & Flax Mills*, Providence, R.I.: Brown Brothers & Co., c1881.

E. Jenckes Manufacturing Co: *Illustrated Price List: E. Jenckes Mfg. Co., Sole Manufacturers of Hicks' U. S. Standard Ring Travelers, Manufacturers' Supplies of Every Description*, Pawtucket, R.I.: E. Jenckes Mfg. Co., 1883.

Lowell Machine Shop: *The Lowell Machine Shop, Lowell, Mass.: Builders of Cotton Machinery of Every Description*, Lowell, Mass.: Stone, Bacheller & Livingston, 1882.

Mason Machine Works: *The Mason Machine Works, Taunton Massachusetts, U.S.A.: Inventors and Builders of Cotton Machinery*, Taunton, Mass.: The Company, 1898.

Merrill, Gilbert R: *American Cotton Handbook: A Practical Reference Book for the Entire Cotton Industry*, New York: American Cotton Handbook Co., c1941.

Scott, Robert: *The Practical Cotton Spinner, and Manufacturer: The Managers', Overlookers' and Mechanics' Companion*, Corr. and enl. Philadelphia: H. C. Baird, 1851.

Webber, Samuel: *Manual of Power for Machines, Shafts, and Belts, with the History of Cotton Manufacture in the United States*, New York: D. Appleton and Company, 1879.

Index

H

K

M

N

P

R

S

T

V

W

Y